CORNWALL'S FUSE WORKS

1831-1961

by

DIANE HODNETT BSc (Econ), MBS

Cover photo: Lighting a fuse at Condurrow Mine, c.1900. J. C. Burrow.

Published by The Trevithick Society
for the study of Cornish industrial archaeology and history

© Diane Hodnett 2016

The Trevithick Society is a Charitable Incorporated Organisation
Registered Charity no. 1159639

Hardback: 978-0-9935021-0-1
Paperback: 978-0-9935021-1-8

All rights reserved. No part of this publication may be reproduced, stored in a retrieval system, or transmitted in any form or by any means, electronic, mechanical, photocopying, recording or otherwise, without the prior permission of the author or the Trevithick Society.

Printed and bound by Short Run Press Limited
Bittern Road, Sowton Industrial Estate, Exeter EX2 7LW

Typeset by Peninsula Projects
c/o PO Box 62, Camborne, Cornwall TR14 7ZN

The Cornish Mine logo is the Trade Mark™ of the Trevithick Society

Clients' calendar for 1900 published by William Bennett and Sons
(Trevithick Society image).

Foreword

This book meets the demand for knowledge of the human aspects of the important industries of the Westcountry. Although Safety Fuse is now of very limited importance to mines and quarries, it was of the greatest value to these industries worldwide until more 'sophisticated' electrical and refined pyrotechnic systems were developed.

The Author has made full use of her access to the Bickford-Smith files; she has given extensive cover to the surprisingly numerous competitors in the Westcountry that emerged. Also, some details of the development of the fuse industry abroad, largely stimulated by Bickford-Smiths, are included.

Technical details have been kept to a minimum, but are well judged to enable an appreciation of their significance on the various developments. Business interests, involvements - and particularly the personalities - are fully treated, as are the harrowing accounts of accidents with explosives; although when the present toll of accidents on the road is considered, these merely pale into a minor footnote,

It is difficult to comprehend now the importance of Bickford, Smith and Davey to the Cornish people. Their operation was of foremost importance. It was an organisation of the first order.

The wonderful multi-tone 'whistle' that used to echo over Tuckingmill to signify shift change, now only haunts the memory bringing back thoughts of a 'world' now, alas, long gone.

Bryan Earl

14th January 2016

Foreword

In 1831 William Bickford invented a fuse that has saved countless lives. There is no statue erected to him, no commemorative plaque in his birthplace in Devon, no honour bestowed by his monarch, and no recognition of his great invention during his lifetime.

This is the story of William Bickford's fuse, and the time when the Camborne and Redruth area of Cornwall was the fuse-making capital of the entire world. The fuse factories are gone now and derelict for the most part. The men and women who worked (and sometimes died) in them should have their part in history recorded. It is hoped that this book goes someway to achieving that.

Bickford's fuse revolutionised safety in the mines. The days of premature explosion of gunpowder underground would become things of the past. Sightless, mutilated miners would no longer be seen, piteously wandering the streets of the great mining towns of Camborne and Redruth, St. Agnes or St. Just, their families condemned to the workhouse. A leather seller from Devon, who had never worked in the tin or copper mines, but driven by his deep Christian faith, determined that he would develop a safe, reliable and portable means of carrying a charge of flame, via a thin rope, towards the gunpowder waiting at the end of the laboriously hammered holes in the rock face, and, crucially, that rate of burning would be a known constant.

He patented his fuse in 1831, and when the patent ran out in 1845, other fuse works were set up in competition to the large Bickford Smith works in Tuckingmill, and Cornish fuse production boomed. When the Cornish mines declined, these smaller fuse works closed. This book is an attempt to record their history also.

Diane Hodnett 2016

Note. While the word 'fuze' may be technically more correct, the fact is that all of the companies involved in the manufacture of the item used the word 'fuse', both in their literature and company names, and therefore the latter spelling has been used throughout.

Acknowledgements

I would like to thank Peter Bickford-Smith for so generously giving me access to his archives, and for providing family photographs for this book.

My thanks must go to the Trevithick Society, whose 80th anniversary was celebrated in 2015. I would like to thank the President Bryan Earl, the Vice-Chairman, Kingsley Rickard, the Publications Secretary, Graham Thorne, and the Curator, Pete Joseph.

I must also thank the staff at the Cornwall Records Office, the Cornish Studies Library and the Royal Institution of Cornwall photographic collection. The permission of the British Library to reproduce sections of William Bickford's 1831 Safety Fuse Patent is gratefully acknowledged. Thanks are due also to: - Barry Bennett, Paddy Bradley, Karen Bickford-Smith, Tony Clarke, Prof Kieran Hodnett, Niamh Hodnett and Duncan Matthews.

I would especially like to thank my family for their support and encouragement – my daughter Nicola, son David and my brother Trevor Andrews.

Contents

PART 1. BICKFORD, SMITH AND COMPANY

1. The early years — 3
2. The Patent — 17
3. The Bickford Smith fuse factory, Tuckingmill 1831-1845 — 20
4. From Tuckingmill to Connecticut – Expansion into the USA — 30
5. Bickford, Smith and Davey – the George Smith years 1846-1868 — 33
6. Bickford, Smith and Co. – the next generation takes charge 1869-1872 — 43
7. Bickford, Smith and Co. – the years of expansion 1873-1888 — 52
8. Bickford, Smith and Co. Ltd. 1889-1899 — 67
9. Bickford, Smith and Co. Ltd. – the Sir George J. Smith years 1900-1921 — 74
10. Bickford, Smith and Co. Ltd – the I.C.I. years and closure 1922-1961 — 102

PART 2. THE CAMBORNE AREA FUSE WORKS

11. The Penhellick Fuse Works, Pool. 1846-1898 — 140
12. William Bennett, Roskear, 1870-1924 — 153

PART 3. THE REDRUTH AREA FUSE WORKS

13. The British and Foreign Safety Fuse Company, Redruth, 1846-1913 — 174
14. Unity Fuse Works, Scorrier, 1846-1918 — 187
15. Tangye's Fuse works, The Elms, Redruth, 1886-1899 — 203

PART 4. A CARADON AREA FUSE WORKS

16. Tremar Coombe, 1855-1870 — 210

List of Figures

1. Bickford, Smith and Co. Ltd. in 1924. The buildings at top right centre were built for munitions work during WWI.   2
2. Ordnance Survey map of Tuckingmill, 1880.   4
3. William Bickford.   5
4. Rope-makers' top.   8
5. Rope-making: The top is being pushed left to right as the rope is being twisted.   8
6. Thomas Davey, 1832.   9
7. Tithe map of Tuckingmill, c.1840 (see figure 8, insert).   10
8. The fuse works at Tuckingmill as shown on the tithe map c.1840.   11
9. A short length of inert fuse.   12
10. A 'pare' of miners hand drilling in a stope, mine unknown.   14
11. The same stope at the end of drilling with fuses lit.   15
12. A paragraph from Bickford's patent. No. 6159, 1831   16
13. Sketch by Luke Herbert   18
14. The needle or pricker.   22
15. Advertisement from *The Mining Journal*.   25
16. Dr George Smith   27
17. Camborne Cross, the site of the present-day library, c.1890.   28
18. Ensign, Bickford & Co. Office, Simsbury, Connecticut   30
19. Souvenir issued by Ensign, Bickford & Co. in 1836.   31
20. Joseph Toy.   32
21. Advertisement in *The Mining Journal*, December 3rd 1853   35
22. Bickford's trademark.   35
23. Blowing up the Vanguard Rock, Plymouth   37
24. *The West Briton*, June 6 1862   38
25. Penlu villas, c 1895. North Roskear Mine is in the background.   39

26. The Smith family, probably at the wedding of William Bickford Smith (centre). 1852	41
27. Cart entering Bickford Smith's factory c 1902	41
28. Tram passing Bickford Smith's factory c 1905. The single storey building on the left was removed c 1912.	42
29. Red River Valley c 1905. Bickford Smith's fuse factory is centre left.	42
30. John Solomon Bickford.	44
31. Advertisement for Bickford, Venning & Co.	53
32. *Post Office Directory* advertisement, 1873.	54
33. Trevarno in the early years, c.1880.	55
34. The Patent Igniter.	55
35. Price list c.1892	56
36. Letterhead from the late 1880s	60
37. The interior of the Porthleven Institute, 2013	61
38. Illustration from the patent application, 1887.	62
39. Illustration from Bickford, Smith and Co's price list.	63
40. Plan to accompany application for an amending licence No. 238, 13th October 1884	64
41. Mrs. Elizabeth Smith.	66
42. Sir George Smith	66
43. William Bickford-Smith	66
44. The factory in 1888. Expansion was possible to the south and west.	68
45. Sir George Smith	70
46. William Bickford-Smith.	72
47. Colonel George Edward Stanley Smith DSO	74
48. Charles E. Tyack	74
49. J. W. Lean	76
50. Plan of Room no. 20. The scene of the fatal explosion at Bickford, Smith and Co. Ltd.	77
51. Bickford, Smith and Co. Ltd Permitted igniter fuse.	79
52. Section through Bickford's Patent 'nippers', with igniter inserted.	80
53. The "BS" "&Co" initials carved on each side of the factory entrance.	85
54. Plan drawn up by ICI Explosives (Ltd) 27/05/1952	
55. The volley firer.	86
56. Bickford Cordeau-Detonant wrapped with a single or two cotton threads	87
57. The lead tube is not covered with thread. From 'Manuel Bickford'	87

58. No. 44 Fuze. CE is 'composition exploding' or 'tetryl'. 88
59. No. 44 Fuze 89
60. Interior of yard building, possible former blacksmiths' building. 1994 95
61. The derelict North Lights building on Pendarves Street. South Crofty Mine behind. 2012 95
62. Bickford, Smith and Co. Ltd.1924. The buildings at top right centre were built for munitions work during WW1. 96
63. Bickford, Smith and Co. Ltd. 1938. The pavilion (1929) (top right centre) can be seen. 97
64. Photograph taken in 1964, not long after the fuse works had closed. 98
65. HRH Prince George on a visit to Bickford, Smith and Co. Ltd. in May 1932. 99
66. The Sports Pavilion at Bickford, Smith and Co. Ltd., Tuckingmill, shortly after its construction in 1929. 99
67. Foremen and forewomen outside the Sports Pavilion. 1933. 100
68. John C. Bickford-Smith and his twin brother William N. Bickford-Smith. 101
69. Harold Octavius Smith 102
70. From a photograph in *The Western Morning News*, December 1923. Note the letter 'A' in 'William' and 'factory'. 103
71. The present plaque (2012). 103
72. The Trelawny Pierrots c.1924. 104
73. Colonel G. E. Stanley Smith presenting the long service awards at Trevarno, July 12, 1928 104
74. A presentation ceremony inside the fuse factory yard. Bickford, Smith and Co. Ltd. 1924. 105
75. A presentation ceremony inside the fuse factory yard. Bickford, Smith and Co. Ltd. 1924. 105
76. Female fuse factory workers. Bickford, Smith and Co. Ltd. 1924. CRO. 106
77. Female fuse factory workers. Bickford, Smith and Co. Ltd. 1924. 107
78. The Welfare Hut 1924 108
79. Coilers 1924 108
80. The employees of Bickford, Smith and Co. Ltd. pictured on the steps of St. Paul's Cathedral on the occasion of their visit to the British and Empire Exhibition, Wembley 1924. 110-111
81. Long-service awards at Trevarno. July 1928. 112
82. Photo of Hayle Towans from The Graphic, May 18, 1901. 113
83. The mess hall and yard c.1930. 114
84. Roscroggan Chapel. Date unknown. 115

85.	Miss Salome Lawrey (centre) and fellow fuse workers at Bickford's. 1928.	116
86.	Miss Winnie Simmons at Bickford's in 1951.	116
87.	Advertisement from *The Cornishman*, 13th June 1946	117
88.	Michael Bickford-Smith	118
89.	Advertisement from *The Cornishman*, 31st August 1950.	118
90.	Bennett & Co. price list, date unknown.	119
91.	Fuse colour samples from Bennett & Co	120
92.	White Countered Gutta Percha Bickford fuse	121
93.	Cordtex.	121
94.	The Bendigo Fuse Factory, Wattle Street, Victoria, Australia.	122
95.	The Institute, Porthleven.	122
96.	Rear of North Lights building.	123
97.	Chapel Road, 2013.	123
98.	The Sports Pavilion 2013	124
99.	Fuse factory façade looking east. 2013.	124
100.	The Bickford, Smith & Co. Ltd. premises and Pendarves Street, Tuckingmill. 2013	125
101.	North Lights building. Pendarves Street. 2013	125
102.	The buildings used for making safety fuse, and metallic fuse. 1994.	126
103.	The buildings where fuses were made are now apartments. 2012.	126
104.	The yard building in 1994.	127
105.	The same building in light industrial use. South Crofty behind. 2012.	127
106.	Derelict building at rear of North Lights, use unknown. South Crofty Mine headframe at rear. 2013.	128
107.	The east and west entrances to the fuse factory. 2013.	128
108.	Former Bennett & Co. offices at Roskear in 1999.	129
109.	Rear of offices, 2012.	129
110.	Bennett's Fuse Works on Roskear Terrace.	130
111.	Cut stone entrance to Bennett's Fuse Works.	130
112.	Rear of Bennett's Fuse Works, 2012.	131
113.	Rear of Bennett's Fuse Works, 2012.	131
114.	Tremar Combe, the Tremar fuse works looking south. 2014.	132
115.	Tremar Combe, the Tremar fuse works looking north.	132
116.	Examples of fuse colours from the Bickford, Smith & Co. Ltd. catalogue.	133
117.	Front page of the long-service awards book.	134
118.	Brunton's Fuse Works on the 1880 Ordnance Survey map.	141

119. Advertisement from *The Mining Journal*, 3rd December 1853. 142
120. The steamer *Indus*. 144
121. Advertisement from *The Royal Cornwall Gazette*. 31st May 1861. 145
122. Advertisement from *The Manchester Weekly Times*, February 19, 1859 147
123. Enlargement of part of the 1924 aerial photo showing the fuse works. 149
124. The Penhellick fuse works is to the left of, and below, the steam train. Taken in 1924. South Crofty in the foreground. 150
125. 1880 Map of William Bennett's Fuse Works, Roskear 154
126. 1907 Map of William Bennett's Fuse Works, Roskear. 156
127. Advertisement in *The West Briton*, June 20, 1889. 156
128. Bennett and Sons advertisement from 1889. 157
129. Nobel advertisement from May 1926. 161
130. The headgear, at the New Dolcoath mine at Roskear, where two were killed through a fall of roof. *Western Morning News*, 7th August 1928. 162
131. The New Dolcoath Mine and old fuse works buildings, 1924. 163
132. Pendarves House. The roof was removed in the mid-1950s. 164
133. Sale notice in *The Western Morning News*, April 9th, 1930 165
134. The two storey building (centre) and the buildings at rear were part of Bennett's fuse works. 168
135. This building backs onto the railway and may have been used as a store by Bennetts. 168
136. This long single storey building backs onto the cricket ground. 169
137. The building used by Bennett's fuse works. 169
138. William Bennett's Fuse Works, Roskear. Looking west. 1924. 170
139. William Bennett's Fuse Works, Roskear. Looking east. 1924. 171
140. Advertisement from *The Royal Cornwall Gazette*, April 24, 1846. 175
141. Tolvean House, Redruth, the home of Alfred Lanyon 176
142. 1880 Map of the British and Foreign Fuse Works. 177
143. Advertisement for The British and Foreign Safety Fuse Co. 178
144. North Redruth, 1880. British and Foreign fuse works and Brewery (right of centre) and The Elms (Tangye's) (above centre). 179
145. Aerial photo (enlarged) from 1924 showing the Redruth Brewery, with the old British and Foreign fuse works to the rear. 181
146. A man working on the stack of the British and Foreign safety fuse works. 183
147. The two young girls are standing on the corner of Rose Hill and Roach's Row, across the road from the fuse works. C 1910. 184
148. View of Redruth, looking north. 1924. Fuse works and Brewery are

	top left. Tangye's (The Elms) is top right.	184
149.	Former fuse works building at Vauxhall, converted to offices for the Redruth Brewery.	185
150.	Fuse works stack at Vauxhall.	185
151.	Advertisement from the *Manchester Courier & Lancaster Advertiser*, May 20, 1848	187-188
152.	Advertisement in *The Freeman's Journal,* Dublin, 12th November, 1858	189
153.	Hawke's engine house, Killifreth, c.1920.	191
154.	Advertisement in the *Western Times*, 10th March 1881	197
155.	1886 Ordnance Survey map of Little Beside and the Unity Fuse Works.	199
156.	Map of 'The Elms' in 1880, the site of Tangye's fuse works.	204
157.	Advertisement for Tangye's fuse, 1898.	205
158.	1906 map showing buildings to the rear of The Elms	206
159.	1924. The Elms (right), with two long buildings to the rear.	206
160.	1883 Ordnance Survey map. The Tremar fuse factory is south of the Chapel, on plot number 131.	212
161.	Advertisement from *The Mining Journal*, August 20, 1853	212
162.	The fuse works building can be seen immediately to the right of the Chapel.	213

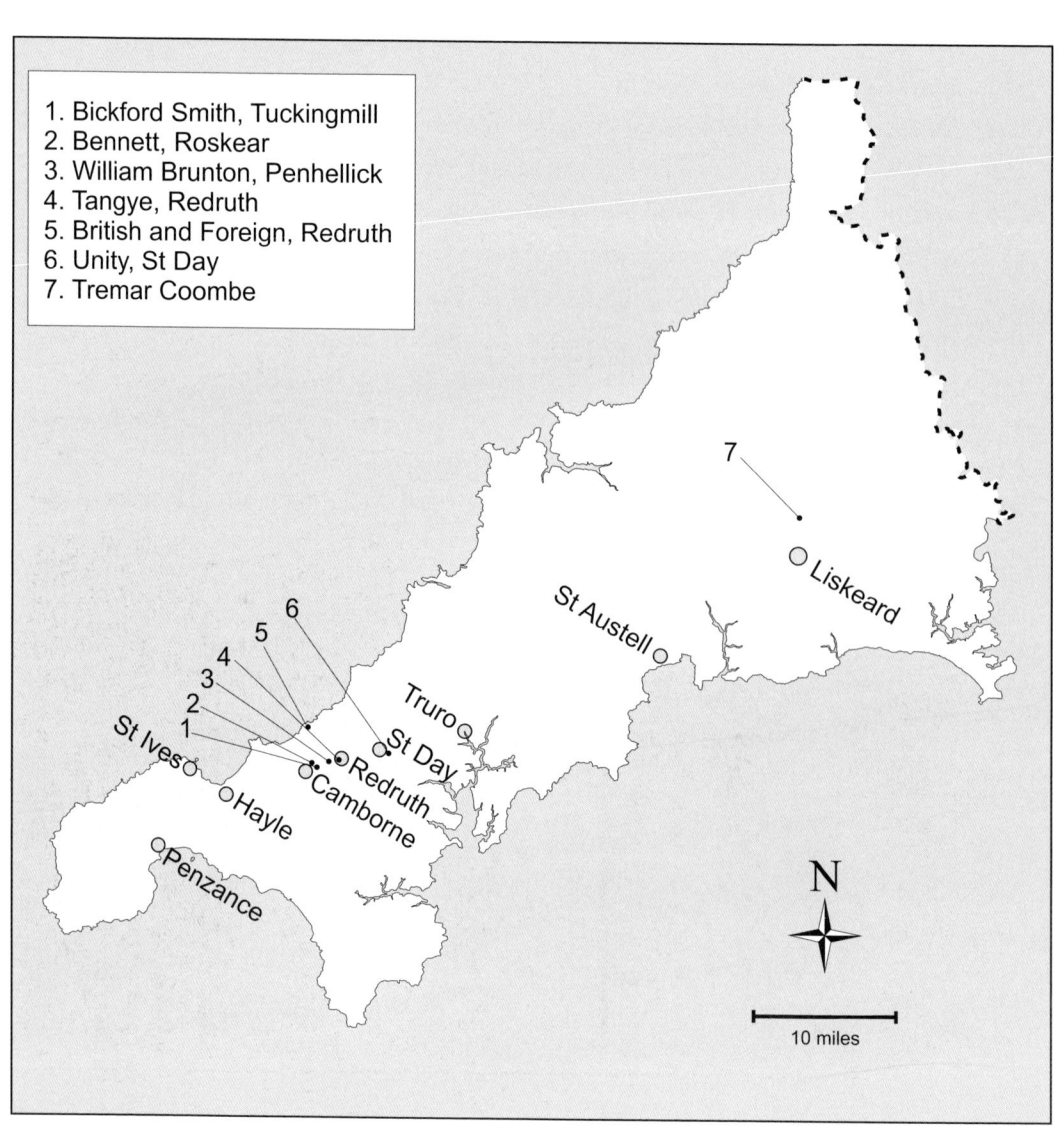

Location map for the Cornish fuse works.

PART 1

THE BICKFORD, SMITH AND COMPANY FUSE WORKS

1831 – 1961

Figure 1. Bickford, Smith and Co. Ltd. in 1924. The buildings at top right centre were built for munitions work during WWI. © English Heritage.

Chapter 1

The Early Years

….at the same moment a loud cry rang through the underground caverns. It was one of those terrible, unmistakable cries which chill the blood and thrill the hearts of those who hear them, telling of some awful catastrophe….The shot had apparently missed fire. Penrose had gone forward to examine it, and it exploded in his face. …..With great labour and difficulty the injured man was half hauled, half carried, and pushed up the shaft, and laid on the grass. 'Is the sun shining?' he asked in a low voice. 'Iss, it do shine right in thee face, Jim,' said one of the miners, brushing away a tear with the back of his hand. So they brought an old door and laid him on it. Six strong men raised it gently on their shoulders, and, with slow steps and downcast faces, they carried the wounded miner home.[1]

R. M. Ballantyne, writing in the middle of the 1860s (but describing a much earlier period) captured the heart-breaking scene where a miner is blinded by an explosion of gunpowder in the blasting hole, after mistakenly approaching the hole thinking that the fuse had gone out. The fuses used to detonate the gunpowder in the Cornish tin and copper mines were of a rudimentary nature. The most common were goose or duck quills, with the feathers shaved off, and pushed one into another to make a long tube. Hollow reeds were also used, and even hollow gorse sticks.

William Bickford (1774-1834), a Devon currier or leather merchant from Bickington, near Ashburton, on the southern edge of Dartmoor, found himself in the heart of the Cornish mining district after his marriage, in Illogan in April 1802, to Susannah Burall, the daughter of a Tuckingmill merchant called Solomon Burall. Four months earlier, on Christmas Eve 1801, Captain Andrew Vivian drove his friend's "puffing devil" up Camborne Hill. Vivian was the friend and partner of the famous engineer Richard Trevithick, and the replica steam carriage is now a well-known spectacle on the streets of Camborne at the end of every April when 'Trevithick Day' is celebrated. It would be interesting to speculate whether or not Bickford stood on the side of the road and witnessed this historic sight.

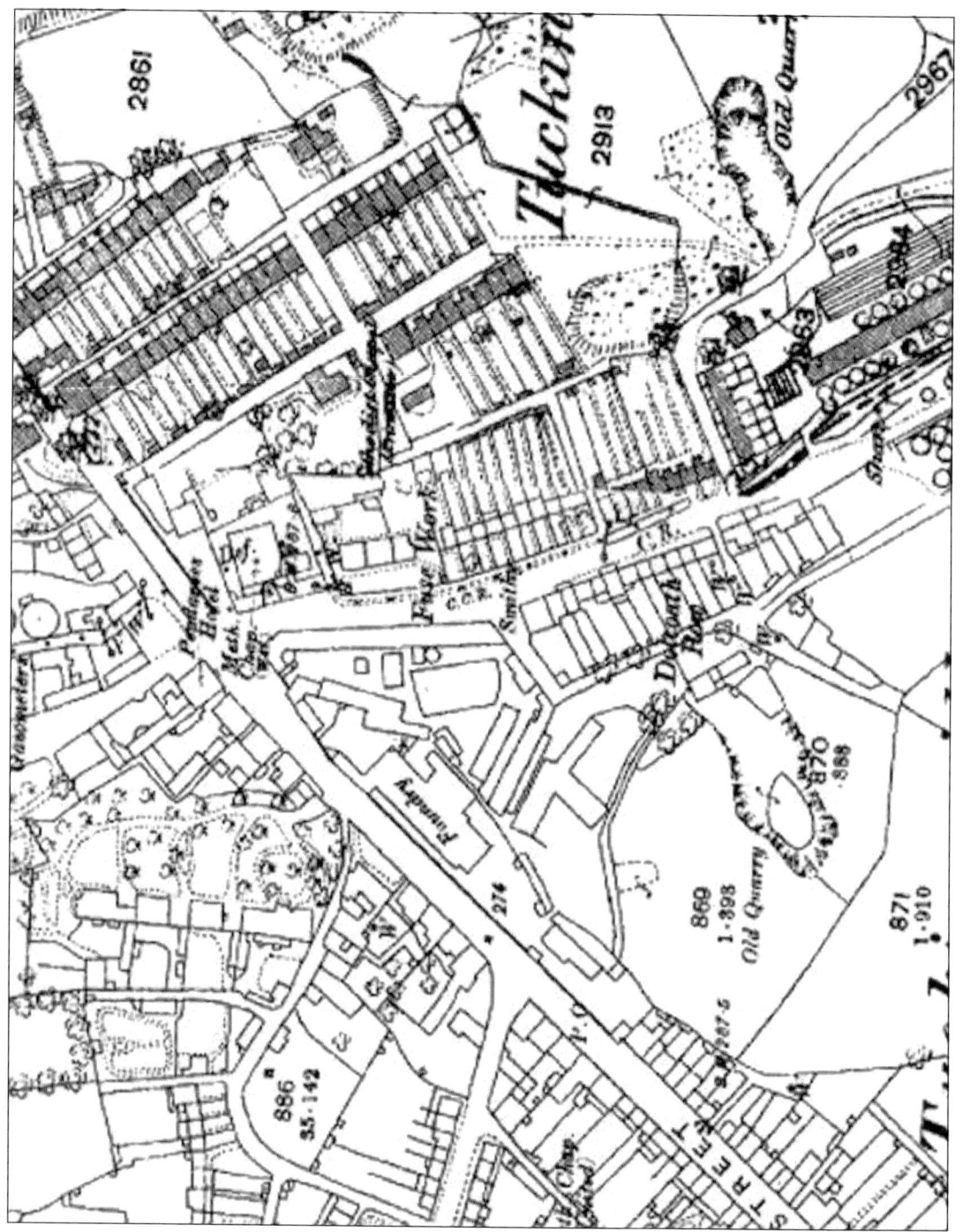

Figure 2. Ordnance Survey map of Tuckingmill, 1880.

Figure 3. William Bickford.

Camborne

The *Universal British Directory* of 1791 carries a good description of Camborne and the surrounding area at that time.

> CAMBORNE Churchtown lies four miles West of Redruth, twelve miles West of Truro, on the great Western road leading from Truro to the Land's End, nine miles North of Helston, eleven miles North-east of Marazion, fourteen miles North-east of Penzance, eleven miles East of St. Ives, all market-towns. It has three fairs annually, viz. March 7, June 29, and November 11; also a show for cattle on Whit-Tuesday. The buildings of Camborne display both uniformity and elegance, the town has the finest land and sea prospects in the county, and

is surrounded by many enclosures of rich and fertile pastures.

About a mile and a half South of the church-town is Pendarves, the family seat of John Stackhouse, Esq. Within the distance of half a mile North are the residences of William Harris, Esq. Josiah Cocke, Esq. and the Rev. John Vivian. Each of these three seats bears the name of Rosewarne.

About a mile South-east of the town is an old copper-mine called Dolcoath, lately stopped from working. It had one perpendicular shaft 172 fathoms, or 344 yards, in depth; the yearly expense of working it, for the last 21 years, amounted, on an average, to £20,000. The adventurers' gains, during that period, amounted in the whole to £80,000 and upwards; and the dues of 1-12th of all the ores, paid to Sir Francis Basset, Bart., the lord, during the same period, amounted to £50,000. About half a mile South-east is another valuable copper-mine, called Wheal Gons and Stray-park, now in working; this mine is on the estates of the Rev. Sir Carew Vivian, Bart, and Sir Francis Basset, Bart.

Between Camborne and Redruth is the parish of Illogan, in which is Tehidy-park, a fine family seat of Sir Francis Basset, Bart. M. P. situate about three miles North-west of Redruth and two miles North of Camborne; the parsonage is an elegant new-built house, one mile East of Tehidy-park, the residence of the Rev. John Basset, rector of Illogan and Camborne, brother of Sir Francis Basset, Bart. The living of Illogan is computed to be worth upwards of £300 per annum, and that of Camborne £500 per annum. In this parish is a copper-mine called Cook's Kitchen, adjoining the Eastern part of Dolcoath, on the estate of Sir Francis Basset, Bart. This mine has long been esteemed the richest in the county.

About a mile and a half West of Redruth, is the village of Pool, through which is the great Western road.

About two miles and a half West of Redruth, and about one mile and a half East of Camborne, on the great Western road, is the village of Tuckingmill. The principal inhabitants are: Paul Burall, Shopkeeper, Solomon Burall, Mercer, and Thomas Hosking, Shopkeeper.'

In 1799 Captain Andrew Vivian, then managing Wheal Gons and Stray Park Mine, took out a lease granted by Francis Basset, 1st Baron de Dunstanville and Basset, for the right to mine for copper from Dolcoath mine, to the east of these two mines, and on the western side of the Red River valley, Tuckingmill. Captain Vivian and Richard Trevithick (junior) held 1/28th and 1/64 share in Wheal Gons and Stray Park Mine, and it very likely that Trevithick held the same small amount in Dolcoath Mine. Captain Andrew Vivian was to be chief mine captain, with Richard Trevithick as chief engineer.[2] It is highly likely that Bickford knew both of these men from distinguished mining families.

Bickford's father-in-law Solomon Burall must have been a very prosperous merchant, for he was one of the adventurers in 1799 in the new Dolcoath mining venture with Captain Vivian and Richard Trevithick. Among the other investors were John Williams of Scorrier, an extremely wealthy mine owner, George and Robert Fox of Falmouth and Perran Wharf, merchants and mine owners, landowner C. B. Agar and John Vivian of Pencalenick, Truro.[3]

After his marriage in 1802 William Bickford moved to Liskeard, where his daughter Susanna was born in 1803 and where his only son John Solomon Bickford was born in 1804, and then moved to Truro, where his daughter Elizabeth was born in 1806. His son was educated at Liskeard, where he attended Mr. Hogg's Grammar School as a day pupil, and was transferred to Dr. Budd's Grammar School, Truro as a boarder when his parents moved to Tuckingmill.[4] William Bickford's wife Susannah died on the 13th of July 1823, and is buried in Illogan with her father Solomon, who died on the 12th of April 1825.

Newspapers such as the *Royal Cornwall Gazette* publicised all mining accidents, and their causes. William Bickford could not have failed to know that many of the deaths were caused by explosions of gunpowder due to very ineffective, primitive 'fuses'. John Ayrton Paris, a doctor as well as Secretary to the Royal Geological Society, Cornwall, was also concerned at the number of disabled Cornish miners. In 1817 he published a booklet entitled: *On the Accidents which occur in the Mines of Cornwall as a consequence of the premature explosion of gunpowder in blasting rocks*. Paris recommended not only the introduction of tools 'which will not strike fire' (*i.e.* spark), but also the invention of an instrument which will 'deliver a charge of gunpowder without loss'. He specifically recommended the introduction of a 'Safety Bar'. The iron bar used for tamping loose bits of rock around the quills used to deliver a charge of gunpowder could generate a fatal spark off these fragments of stone. Paris knew this could be solved by adding a non-sparking alloy of tin and copper to the tip. He added the information that this 'Safety Bar' had been in general use for more than a year in the mines of west Cornwall.[5] It was to be another thirteen years before his second recommendation was met in full by William Bickford's 'Safety Rod'.

It is believed that William Bickford thought that it might be possible to insert a fine stream of gunpowder thought the centre of a narrow rope while watching James Bray making rope in his ropewalk, and thus fabricate a much safer fuse. Very little is known about James Bray, however. He was a Steward of the Wesleyan Society at Tuckingmill, and according to Bickford's son-in-law was 'a very respectable man'.[6]

In the 1841 Tuckingmill census, James Bray, 57, a rope-maker, is listed as living on Illogan Downs with his wife Francis, daughter Mary and son John (also a rope-maker). Not far away lived rope-makers Peter Knight and Edmond Spargoe. The 1840 Tithe Map for Illogan shows James Bray leasing six acres on the road to the

Eastern Lodge of the Tehidy Estate, the home of Lady Basset. A mile further northeast, Thomas Hitchens is the landowner of 'ropewalk meadow', Fairfield. This ropewalk leading to Trengove is marked on the Tithe map.[7]

Ropewalks are so-called because, in order to twist a rope, the individual strands have to be made first. In order to do this a man would fasten a bundle of jute or hemp around his waist, or put it in a large bag at his waist, and then, slowly walking backwards, would pay out the fibre into the twisting yarn as he did so. The end he was facing was fastened to a hook in the middle of a wheel which his assistant spun. Each piece of fibre was added and twisted into the spinning yarn. When three or more strands of the desired length had been spun, they were twisted together to make a rope.

The three strands were kept separate by using a rope-makers top, tapered at one end to feed the strands into the spinning rope.

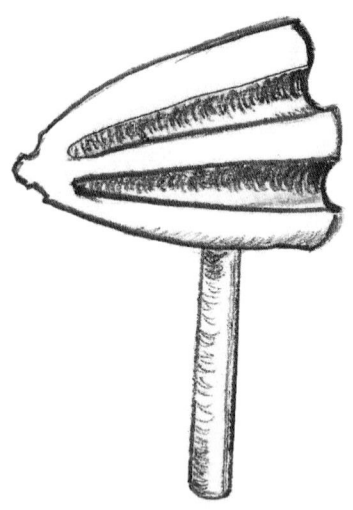

Figures 4. Rope-makers' top.

William Bickford understood that if this process was turned from a horizontal process to a vertical process, it should be possible to trickle a small stream of gunpowder into the centre of the rope by substituting a funnel in place of the rope-makers top. Like all brilliant inventions, turning this into reality was another matter. He turned to a friend, Thomas Davey, who was clearly a very practical man. Davey was a working miner and a Methodist Class Leader. "To Mr Thomas Davey, of Tuckingmill, belongs the larger share of the credit for the original mechanical appliances". So wrote Bickford's grandson, Sir George Smith, in 1909.[8]

Figures 5. Rope-making: The top is being pushed left to right as the rope is being twisted. Photo taken at Morwellham Quay, Devon, by the author.

Bickford's son-in-law, Dr George Smith, writing in his autobiography, describes the trials that Davey and Bickford undertook before the fuse could be made to their satisfaction. They found that the stream of gunpowder leaked from the unspun fibres of hemp or flax. After numerous trials they tried flax yarns, and found that a sufficient number, placed at the neck of the gunpowder-filled funnel, and twisted as they were drawn out, retained the stream of gunpowder. They then devised a method of winding another series of cord around the gunpowder filler core, in order to make it quite firm. It was then passed through tar varnish. Dr George Smith added: - 'I had marked every progressive step in this invention, and although I had at the time no idea of its being of much pecuniary value to the inventor, I had a very sanguine opinion of its vast superiority in point of safety to everything previously used for the purpose of conveying fire to the charge in blasting'.[9]

The earliest place of manufacture was close to the Red River, on what is now Chapel Row (then known as Dolcoath Row). The existing fuse factory buildings on the southern side of Pendarves Street were built around 1882.

The earliest map of Tuckingmill is the Tithe map of 1840, and it shows a small hamlet. It can be seen that the northern side of Pendarves Street was built by 1840. This street of workers' houses was built by the Pendarves landowners on what was then good agricultural land. The land on the other side of the Red River was owned by the Basset family, who built rows of houses on what was old mine waste.

Figure 6. Thomas Davey, 1832.

It is interesting to look at this Tithe map in some detail. On the plots numbered

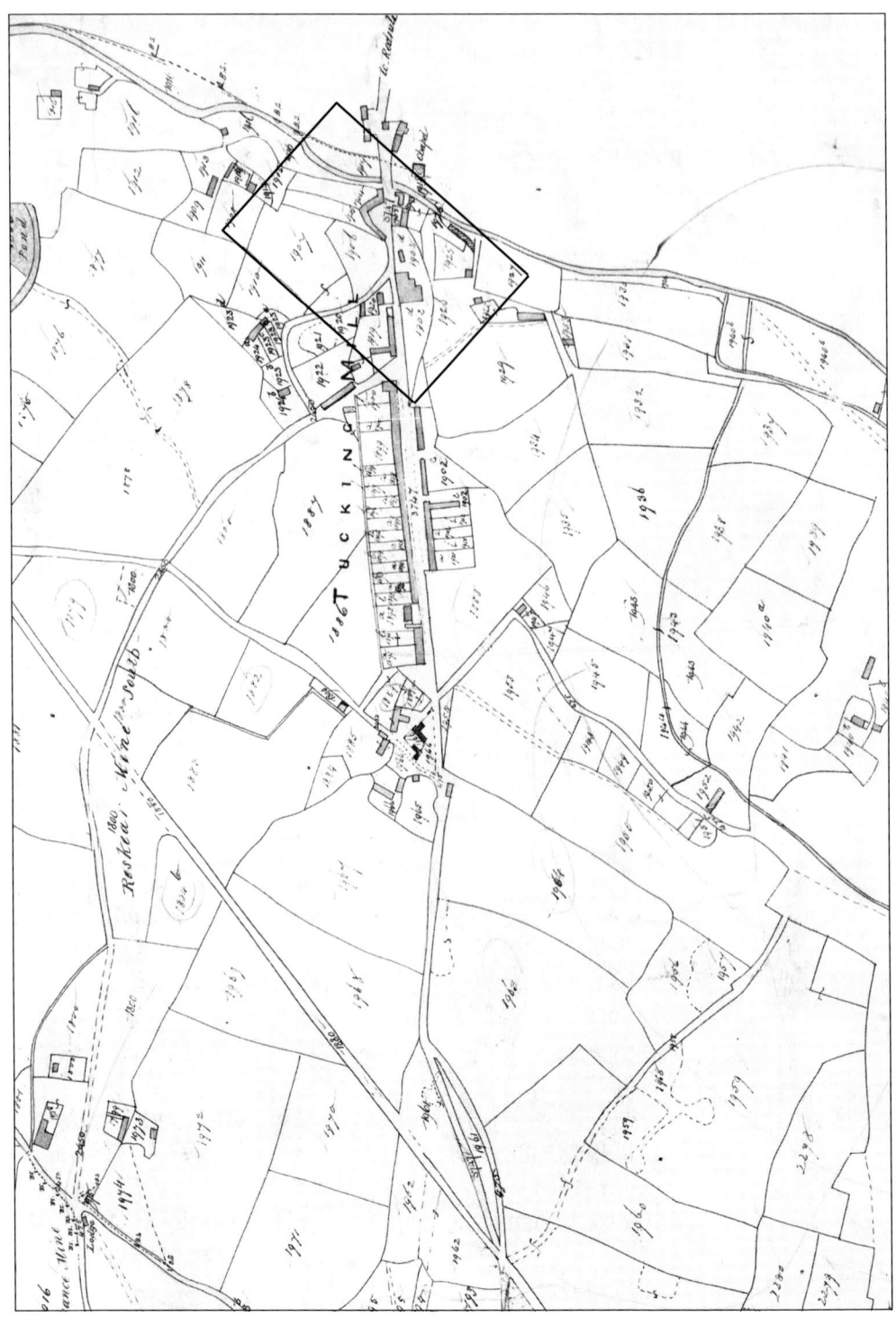

Figure 7. Tithe map of Tuckingmill, c.1840 (see figure 8, insert).

1904 and 1905, the lessee is Grace Burrall, with the occupier listed as 'herself'. The plots each contained a house and an orchard. The Camborne Gas Company was later established on plot 1904, which is now a park. Plot 1917 is recorded as leased by Kitty Burrall, on the east side of the Red River, with the occupiers named as Thomas Mitchell and John Eddy, and containing houses and gardens. Plots 1906 to1912 were leased by Mary Gribble, who appears in the 1841 census as an innkeeper.[10]

On the south side of the turnpike road, Pendarves Street, plots 1903 and 1925 were leased by Grace Burrall from the Honourable Lady Basset. While plot 1903, fronting the new road, were occupied by Grace Burrall, and contained houses and gardens, plot 1925 is sublet to Bickford and Co., and contained a "Safety Rod Manufactory". The relationship between Grace Burrall

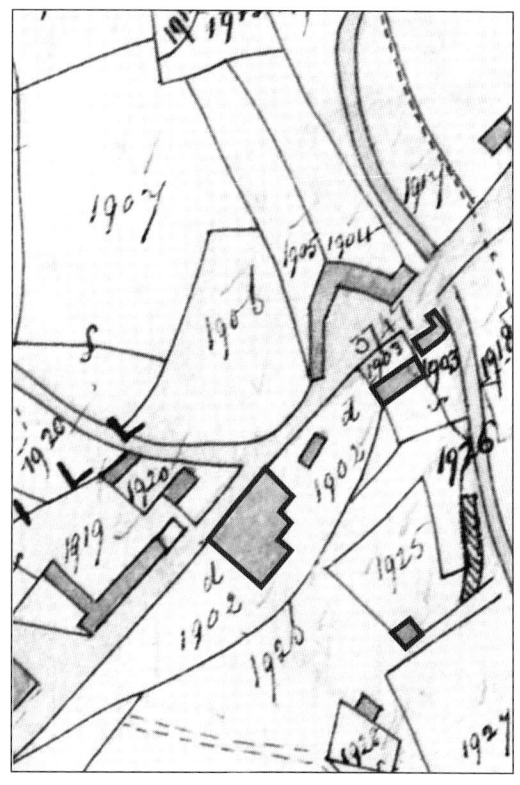

Figure 8. The fuse works at Tuckingmill as shown on the tithe map c.1840.

and William Bickford's wife, Susannah Burall, is not known. The long building drawn running northeast-southwest on this plot is heavily cross-hatched. This may indicate that it was either in the process of being built, or was derelict. Plot 1926 was owned by Lady Basset. According to A. K. Hamilton Jenkin, the workings of the old copper mines known as Wheal Plosh and Wheal Susan lay beneath the fuse factory. In 1760 the Wheal Plosh sett was extended eastwards to encompass the Longclose Lode, and also to take in 50 fathoms of the Copper Tankard Lode to the south of Longclose. From 1763 to 1773, records show that 400 tons of copper ore from this source were sold for £2,800.[11] Wheal Susan was south of the main road, on the eastern side of the valley.[12]

Plot 1902d, close to the fuse factory, was leased by William Vivian from E. W. W. Pendarves, and is listed as containing a Foundry, Yard, Pond, Houses and Garden. Vivian was a former employee of the Copperhouse Foundry. The foundry was established in 1832 by Vivian, who acquired the Roseworthy Hammer Mills about ten years later. They were owned at that time by his cousins John and Joseph Vivian.[13] The Tuckingmill Foundry was taken over by the fuse factory in 1910, when land to the south-west of the foundry became the site of the North Lights building, which was used as a jute spinning works. William Vivian also leased two adjoining plots directly across the Pendarves road from his foundry (plot numbered 1920 and 1921), listed as 'House, garden and Plot' and 'Wastrel' respectively.[14]

There were two large foundries in the vicinity of Tuckingmill at this time, supplying goods to the expanding mines. By the middle of the eighteenth century, a small brass foundry was in production on the site of an old tucking (fulling) mill, and powered by water from the Red River. It was taken over by 1793 by John Budge, who was chief engineer at Dolcoath Mine. Another iron foundry, operating by 1806, stood close to the end of the present day Dolcoath Avenue, where it appears in records as 'Dolcoath Foundery'.[15] In addition, there would have been a number of blacksmiths working in small premises, producing not only items for the mines, but products for farmers and for domestic use.

The technology to spin thread was already in existence. In 1770 James Hargreaves

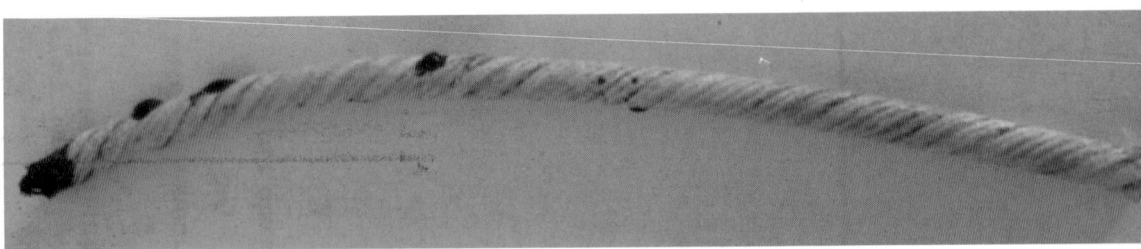

Figure 9. A short length of inert fuse.

had invented what became known as the Spinning Jenny. The machine used eight spindles onto which thread was spun by turning a single wheel by hand. Further improvements increased the number to eighty. The thread produced, however, was coarse. Richard Arkwright, a wigmaker from Preston in Lancashire, together with clockmaker John Kay and other craftsmen, developed a machine that used three sets of paired rollers that turned at different speeds. A set of spindles twisted these fibres firmly together, thereby producing a stronger yarn. This was patented in 1775. Arkwright was the first to use a Boulton and Watt steam engine to power his textile mills.

Bickford's idea was to use spindles or bobbins to weave a strong thin cord around a core of gunpowder – a far more dangerous and complicated operation than merely spinning thread to be made into textiles. In addition, because of the wet mines, this cord had to be made waterproof in order to keep the gunpowder core dry.

It is not known where William Bickford and his partner Thomas Davey actually made their prototype spinning machine. It may have been made in the small building shown in the corner of Plot 1925 on the Tithe map, or it may have been made in a nearby foundry or blacksmith's shop. In 1832, William Bickford wrote of 'much labour and expense' in adapting his fuse to 'popular use'. He refers to a "course of experiments with rock blasting". Eventually reliable fuse of a uniform strength and thickness was produced. Local miners were given fuse to test.[16] All that remained was to patent the invention, and persuade the Cornish mines to use it.

References, Chapter 1

1. R. M. Ballantyne. *Deep Down. A Tale of the Cornish Mines.* First published 1868.
2. Allen Buckley. *Dolcoath Mine. A History.* The Trevithick Society. 2010
3. *Ibid.*
4. The *West Briton and Royal Cornwall Advertiser*, Thursday evening. February 24 1870.
5. John Ayrton Paris. *On the Accidents which occur in the mines of Cornwall. 1817. Penzance.* Printed and sold by T. Vigurs.
6. The autobiography of George Smith. LL.D . Cornish Studies Library, Redruth
7. Tithe Map. Cornwall Record Office.
8. Sir George Smith. The *Rise and Progress of the British Explosives Industry,* 1909. (ebook)
9. The autobiography of George Smith. LL. D. Cornish Studies Library, Redruth
10. Tithe Map. Cornwall Record Office.
11. A. K. Hamilton Jenkin. *Mines and Miners of Cornwall.* Volume 10, Camborne, p 15.
12. T. A. Morrison. *Cornwall's Central Mines. The Northern District. 1810-1895.* p 256
13. http://www.mocavo.com/A-History-of-the-County-of-Cornwall-Volume-1/111541/66111
14. Tithe Map. Cornwall Record Office
15. Allen Buckley. *Dolcoath Mine. A History.* The Trevithick Society. 2010
16. The autobiography of George Smith. LL.D. Cornish Studies Library, Redruth

Figure 10. A 'pare' of miners hand drilling in a stope, mine unknown.
Trevithick Society archive.

Figure 11. The same stope at the end of drilling with fuses lit.
Trevithick Society archive.

NOW KNOW YE that in compliance with the said recited proviso I, the said William Bickford, by this instrument in writing under my hand and seal, and by the Drawings hereunto annexed, and the references thereto, and the explanations and descriptions therein and herein-after mentioned and contained, do particularly describe and ascertain the nature of my said Invention, and in what manner the same is to be performed as follows, that is to say:

The instrument invented by me for igniting gunpowder when used in the operation of blasting of rocks and in mining, which I call "the miner's safety fuze," I manufacture by the aid of machinery and otherwise of flax, hemp, or cotton, or other suitable materials spun, twisted, and countered, and otherwise treated in the manner of twine spinning and cord making, and by the several operations herein-after, and in and by the Drawings hereunto annexed, mentioned and described, by means whereof I embrace in the centre of my fuze, in a continuous line throughput its whole length, a small portion or compressed cylinder or rod of gunpowder, or other proper combustible matter prepared in the usual pyrotechnical manner of firework for the discharging of ordnance, and which fuze so prepared I afterwards more effectually secure and defend by a covering of strong twine made of similar material, and wound thereon at nearly right angles to the former twist by the operation which I call countering, herein-after described, and I then immerse them in a bath of heated varnish, and add to them afterwards a coat of whiting, bran, or other suitable powdery substance to prevent them sticking together, or to the fingers of those who handle them; and I thereby also defend them from wet or moisture or other deterioration, and I cut off the same fuze in such lengths as occasion may require for use. Each of these lengths constituting, when so cut off, a fuze for blasting of rocks and mining, and I use them either under water or on land, in quarries of stone and mines for detaching portions of rocks, or stone, or mine, as occasions require, in the manner long practised by and well known to miners and blasters of rocks.

Figure 12. A paragraph from Bickford's patent. No. 6159, 1831 (British Library)

Chapter 2

The Patent

To W. Bickford, of Tucking Mill, in the county of Cornwall, Leather Seller, "for an instrument for igniting gunpowder when used for blasting rocks, which he denominates the 'Miner's Safety Fuse'", a patent was granted on the 6th of September, 1831, and the specification was lodged in the Enrolment Office on the 28th of February, 1832.

In *The Register of Arts and Journal of patent inventions* of April 1832, from which the above quote is taken, the editor Luke Herbert wrote this about Bickford's patent: 'the patentee verbally describes, with great attention to exactness, the most minute and unimportant parts of his very simple apparatus, which are moreover fully explained by elaborate and correct, though coarsely executed, drawings; and so liberal has Mr Bickford been to the public, in return for the exclusive privilege of the patent, that he indulges them not only with the representation of his mechanisms, but every door, window, wall, board, joist, rafter, roof, and almost every nail of his extensive manufactory is placed before them in every point of view for examination; thus fulfilling the law not merely in the spirit, but beyond the letter thereof'.[1]

Herbert provided his own sketch of the yarn twisting process, copied from Bickford's patent. The drawings in Bickford's own specifications are exceptionally detailed, as are the accompanying descriptions. This may explain why, when the patent expired in 1845, competitors in Cornwall were able to very quickly set up their own fuse factories.

There was to be a challenge to Bickford's patent, which took place after William Bickford's death in 1834, when the factory was under the charge of his son-in-law.

The Legal Challenge to Bickford's Patent, 1839
In 1837, the firm of Bickford and Co. took out an injunction to prevent a gentleman by the name of 'Skewes, the Younger' infringing their patent by selling his own fuses. The case was heard before a judge and jury at Devon Assizes, in Exeter, on July 29th, 1839. The jury were to be selected from the east of the city, where there was no mining.

A fuse was produced in court, and the manufacturing process was described. It was explained that a woman named Hosking, employed by Bickford and Company, went to work for Skewes in 1837. Skewes built a machine resembling Bickford's, and started to sell fuse.

Skewes alleged that William Bickford was not the original inventor, but that the safety fuse was in public use before he took out his patent. He stated that he had not infringed the patent, as neither the machine, the material of the fuse, or the fuse itself, were not properly described in the patent, as required by law.

A report of the time stated that a large number of witnesses were examined on both sides, but they 'did not bear out the Cornish motto of "One and All", as they told different stories'. Eventually the jury returned their verdict in favour of Bickford and Company, who were awarded damages of one shilling.[2]

Dr George Smith (Bickford's son-in-law) explained the whole case more fully in his autobiography. Smith wrote that Skewes, a carpenter, occupied part of an unoccupied malt house. He became very attentive to one of the girls making fuses for Bickford, and persuaded her to work for him. She, of course, took with her all her information regarding the machinery and manufacture of Bickford's fuse.

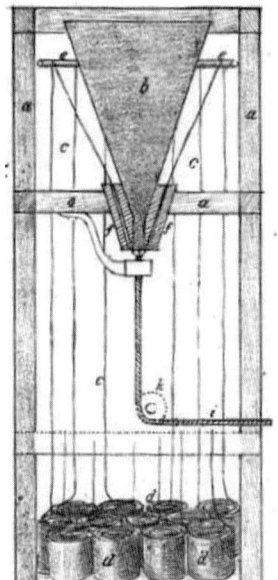

The company solicitors in London advised Smith to provide proof that Skewes had actually sold fuse. This proved difficult – 'he always worked with the doors locked and refused admission to everyone' according to Smith. Eventually proof was obtained from one or two commercial travellers. An injunction was prepared, and served on Skewes with difficulty:- 'after midnight....as he was going to his father's house'. Skewes replied that he was going to court to have the injunction removed, and George Smith wrote that he had to visit London at least five times to prevent this.

The trial began in the summer of 1839. Smith and his

Figure 13. Sketch by Luke Herbert

partner Thomas Davey brought about twenty witnesses with them, together with several from the nearby Tavistock district. A Colonel Pasley was called from Chatham (where there was a Royal Naval dockyard), a Colonel Jackson from London, Mr John Budge ('the most respected Quaker in Camborne') and the principal mine agents from Camborne.

One of the fuse factory workers, named as Mary Davey, was called to give evidence, and it is from her that some idea of the factory may be gleaned.

> I am in the employ of the plaintiffs, at Tuckingmill, in Camborne. We have store houses, steam engines, and powder magazines, and about 32 persons all employed at this manufactory. When I first came to the factory the fuses were made by a machine similar to this. Within the last two years an alteration has been made in the reels of thread, which are now put on top instead of under.

Smith wrote that about forty of the witnesses, together with himself and Thomas Davey, lodged at the Half Moon Hotel in Exeter. Before dinner, he was given legal advice that he should prevent his witnesses from walking outside, in case they met opposition witnesses, and 'unadvised conversations' took place. In order to keep them in the hotel, Smith gave a two hour lecture after dinner on 'The progress of mechanical invention with special reference to the safety fuse'. The next day, a Sunday, he preached in St. Sidwell's Wesleyan Chapel. On the Monday, the trial was resumed, which lasted all day, at the end of which a verdict in Bickford and Co.'s favour was reached.

The morning after the trial Skewes' father approached Smith, and asked him for a loan to take him and his witnesses back to Camborne. George Smith declined to give it.[3]

The verdict copper-fastened Bickford and Co.'s right to the monopoly on fuse-making. In his autobiography, Smith noted: - 'from that time the fuse trade regularly increased for a considerable time'.

References, Chapter 2

1. *The Register of Arts and Journal of Patent Inventions.* Vol. 7. 1832. Edited by L. Herbert. (ebook)
2. *The Western Times*, Exeter. Saturday August 3rd 1839. The case was widely reported at the time.
3. The autobiography of George Smith. LL.D. Cornish Studies Library, Redruth

Chapter 3

The Bickford Smith Fuse Factory, Tuckingmill
1831-1845

In September 1831 William Bickford was granted his 'fuze' patent. It is interesting that the patent was in his name only. His partner, Thomas Davey was not mentioned. He was, however, to receive a quarter of the profits.

A very early invoice, dated 11th October 1831, is made out to South Roskear Mine. The coils were 30 feet long and cost one shilling and three pence each. The invoice carried the written heading 'Bickford and Davey'. By 1835, the invoices carried the printed heading 'Bickford, Smith and Davey'. By the end of 1833, coils were 24 feet or 48 feet and the price was one shilling (twelve pence) for 24 feet. During 1834 the price came down to nine and a half pence and by 1837 it was reduced further to eight and a half pence. It is believed production in 1833 was about 40,000 coils.[1]

William Bickford worked to make a waterproof cover of caoutchouc (a natural rubber also known as India rubber) for his fuse. A Mr. Hancock of London, who had been involved with Bickford in the experiment with developing caoutchouc tubes, was offered a special sales agency at 25% discount, when the usual discount was 10%. It was feared that he might develop a competitive fuse, namely caoutchouc tubes filled with gunpowder.

Around 1836 a Mr. Hancock from the Hayle area, succeeded in making a braided fuse, with more expensive twine than Bickford's spun fuse. Hancock could not compete on price, however, and eventually his machinery was purchased by Bickford, Smith and Davey, and removed to the Tuckingmill factory. It is not known if this Mr. Hancock was the same man who had worked with William Bickford on the caoutchouc experiments.[2]

Four years earlier, Bickford's younger daughter Elizabeth had married George Smith, a carpenter and lay Wesleyan preacher from Condurrow, a mile south of Camborne. He was to play a pivotal role in the expansion of the fuse business.

Writing about the early years of the fuse business, Smith described how these 'cords' were given out early in 1831 for trial by miners in neighbouring mines. Where they were used according to the instructions given to the miners, they were found to work extremely well.[3] As is the case with a lot of new inventions, ignorance and prejudice prevented a widespread acceptance. This particularly applied to some of the mine managers, pursers and their relations, who had a lucrative trade supplying the old goose quill 'fuses'. From the beginning, the fuse was cut into lengths of 24 feet, and burnt at the rate of 30 seconds per foot (1 second per centimetre).

Bickford himself travelled around the local mines after the patent was granted, and persuaded some of them to give him orders. Eventually, he decided to produce a small pamphlet explaining the advantages of the new fuse. George Smith, not satisfied with Bickford's endeavour, re-wrote it, and it was circulated to the local mines.[4] It was printed in 1832 by L. Newton, bookbinders of Camborne, and was entitled: - 'The Safety Fuze or an appeal to practical miners on the utility and general advantage of "THE MINER'S SAFETY ROD" (so termed in Cornwall) for Blasting Rocks etc'.[5]

In this booklet, Bickford discussed the merits of his fuse under four headings: - safety, economy, speed of despatch (use), and simplicity.

Safety
Writing in 1832, Bickford noted that his fuses had 'long been used extensively' in the nearby North Roskear Mine. He noted that the agents there kept accurate accounts of failures, and in the last two reports (of two months each) they had reported 1 in 880 failures and 1 in 849 failures of his fuse.

He wrote that he was surprised that any failed, but he knew miners used fuse to tie up tools to be sent underground, and also tied up old shoes – and subsequently used the fuse for blasting. He had also learnt that fuse was left underground for a week or two in very damp conditions.

He compared the failure rate in fuses with that of quills, and pointed out that 'impartial and intelligent miners' gave their opinion on the average number of failures that occurred in the use of quills. Some estimated five in every hundred failed; others thought four in every hundred failed.

Economy
Bickford wrote that if fuse was to be sold at one halfpenny per foot, some users might think that it would be cheaper to use the needle and quill method. However, he pointed out that less gunpowder would be needed in blasting, which would off-set the cost of the fuse. To do this he needed to compare the quantity of powder used when blasting using fuse and using needles and quills.

He described how he carried out a series of tests or 'experiments' with rock blasting, but found it impossible to procure a number of holes which provided equal resistance to the action of the powder. This made it impossible to measure the increase or decrease in power. The location of these blasting tests was not named by Bickford, but the nearby quarries on the northern flanks of Carn Brea could have been used.

He further experimented to find the loss of explosive power when using the needle or 'nail', compared to the use of the fuse in blasting. To this end he suspended a gun-barrel by two cords, and measured the recoils when a measured amount of gunpowder was used. He enlarged the touch-hole for a second round of tests and found a waste of gunpowder by about 15 percent. He pointed out in the booklet that the fuse could never fill an aperture greater than one eighth of an inch, which was the size of the column of gunpowder in it. The needle by comparison was larger than the fuse at the end which entered the gunpowder, and usually measured a quarter of an inch or more, and it was much bigger at the other end. He described how, if the tamping was not very firm, and the rock was very hard and resistant, power could momentarily escape and enlarge the bottom of the needle-hole, which lead to an increasing escape and therefore loss of power. He fixed the loss of power in this way, using the needle, as 20 -25%. He compared this with the minimum loss of explosive power if his fuse was used. He also stated that quills were also larger than the fuse.

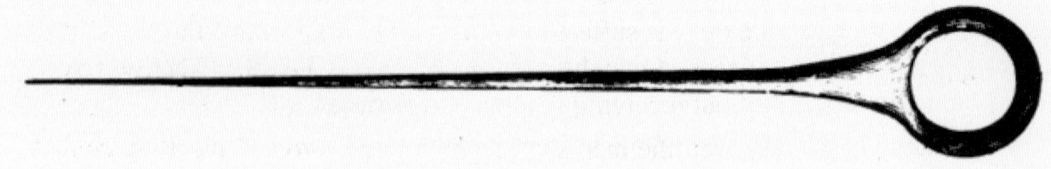

Figure 14. The needle or pricker.

He also thought that as the fire proceeded along the length of a lit fuse, removing the powder, the heat restored the elasticity of the varnish coating and the fibrous body, which lead to the closing of the hole occupied by the fuse, allowing no opening whatsoever for the escape of any part of the explosive power.

Bickford quoted 'accurate calculations made by some experienced mine agents'. These calculations showed that the fuse, at a halfpenny per foot, amounted to the same as quills which sold for 3s 3d (39 pence) per thousand and powder at 45s per hundredweight (112 pounds) He thought that, taking into account the greater number of quills that failed, a saving of 4% could be used by using the fuse.

In the booklet, Bickford also asserted that it was well known that if a measured charge of gunpowder is 'exactly proportioned' to the resistance of the rock, without any allowance for this waste of power, in many cases the rock was not blasted at all, or the tamping moved, as " the entire power of the charge escapes through the nail-hole." He claimed that, because allowance has to be made for an obvious waste of gunpowder, using the fuse was "not only safer but cheaper than the nail in blasting."

Despatch
Bickford compared the speed of use of the fuse, compared to using the needle and quills. He noted that the fuse just had to be cut to length and 'the certain result is to

leave the effects of the powder on the rock the greatest possible'.

Simplicity
Bickford explained that his fuse was manufactured in lengths of 60 feet, was extremely portable, and could be cut to any length, inserted into the hole and lit, where it conducted the fire to the charge 'slowly and surely'.

He stated that the fuse was a necessity in underwater blasting. He wrote that it was certain from 'actual experiments' that it would lead the fire to a charge of powder to a depth of six or seven fathoms. Unfortunately he did not elaborate on the location of these experiments.

Bickford concluded his 28 page booklet with *Directions for using the Fuze*.

1. Let it be used solely for the purpose of blasting, and not to bind up tools, or to serve instead of cord.
2. Let it be kept in a place that is tolerably dry until it is wanted for use. It is believed, that if the Fuze is suffered to remain any considerable time in very damp or mouldy places, it may be injured; although it has been kept several weeks in water without receiving any sensible damage.
3. If the mine is very wet, the men should ask for sump-rods from the person appointed to deliver the materials, as these are intended expressly for this purpose.
4. There should in all cases be some powder put into the hole before the Fuze, and in very wet places it may be necessary to make a hole through that part of it which will stand in the centre of the charge, that the fire may certainly be communicated to dry powder.
5. Before the Fuze is placed in the hole the outside or countering thread should be stripped down about an inch from the end which will be uppermost. This not only prepares it for taking fire more readily; but also consider ably lessens the quantity of smoke, which is a great advantage in close spaces. The Fuze requires rather more than half a minute to burn a foot in length in the open air, but may possibly take a minute a foot in a close hole.
6. The miner should be careful to use a proper substance for tamping, and as that which is soft and tenacious will floor best, it should always be preferred.

It is interesting that Bickford wrote that his fuse was made in lengths of 60 feet. It is generally accepted that it was sold in lengths of 24 feet.

In May 1833, Bickford and Davey felt compelled to write a letter to the *Royal Cornwall Gazette*. The letter stated, among other items, that they had often heard of miners binding up their tools and tying on their shoes and using the fuse for 'other purposes' before it is used for blasting.[6] They asked that the fuse be used solely for blasting. They noted that such 'abuses' were more likely to prevent the fuse from conveying fire to the charge. However, well into the 20th century, Cornish miners were using lengths of fuse as straps to tie their trousers below the knee. These straps were known as 'yorks' and prevented water and mud from getting up the trouser legs.

North Roskear Mine was situated on the side of the old main road via Scowbuds from Tuckingmill to Camborne. It was a very heavy user of the new fuses from the start, mainly because it was expanding rapidly, having hit ore of a superior grade in 1829, and by 1831 it was making a profit of £600 a month. The first report from there noted one failure in 750 fuses, the second report noted one failure in 800 fuses, and the third report recorded one failure in 1,533 fuses.[7] It appears that Bickford and Davey had managed to produce a consistently reliable fuse right from the beginning.

Towards the end of 1832 William Bickford was taken ill, suffering paralysis, loss of speech and dementia. Bickford's only son, John Solomon Bickford, was at that time a school master in Hayle. Not knowing William Bickford's wishes regarding the growing fuse business, the family consulted his will.[8] They discovered that his eldest daughter Susanna was to receive an annual annuity of twenty pounds, and his son John Solomon and daughter Elizabeth Smith were to receive all Bickford's other property. The will stipulated that the fuse factory management was to be left to George Smith, who was to be paid one hundred pounds per year 'so long as he continues faithfully, carefully and diligently to act thereon'.[9] Smith wrote later that Thomas Davey (who owned one quarter of the fuse business) was unhappy with this decision, while Bickford's daughter Susanna ('never of an amiable disposition') was also extremely unhappy with the terms of her father's will.

Smith closed his carpentry business, and became the full time manager, with Thomas Davey, of the Tuckingmill fuse factory.

One of the first things he had to deal with was William Bickford's share in Relistian Mine, Gwinear, which became bankrupt with debts of many thousands of pounds. Smith eventually paid Bickford's share of this debt, amounting to between one thousand and twelve hundred pounds. This was a large amount for a small but growing firm, but Smith acknowledged that, in the course of dealing with it, he came into useful contact with some of the principal bankers and merchants of West Cornwall.

William Bickford died in October 1834. In his autobiography, Smith wrote:- 'this event, by releasing me from such constant contact with my sister-in-law, was a great relief'. The fuse business was growing and required Thomas Davey's full time presence in Tuckingmill, while Smith visited the mines of Cornwall and West Devon looking for orders. The fuse was recommended to the Board of Ordnance by Colonel (later Sir Charles) Pasley. They used it to great effect blasting under water to deepen Kingstown (now Dun Laoghaire) Harbour, Dublin, in the 1830s and 1840s. This underwater use in Ireland was one of the first applications of the use of gutta percha to make the fuse waterproof. This covering is the gum of a tree native to Malaysia. Because gutta percha degrades in hot climates when exposed to the air, Bickford, Smith and Davey developed an extra coating to cover the gutta percha. This was originally produced for the Indian Government.

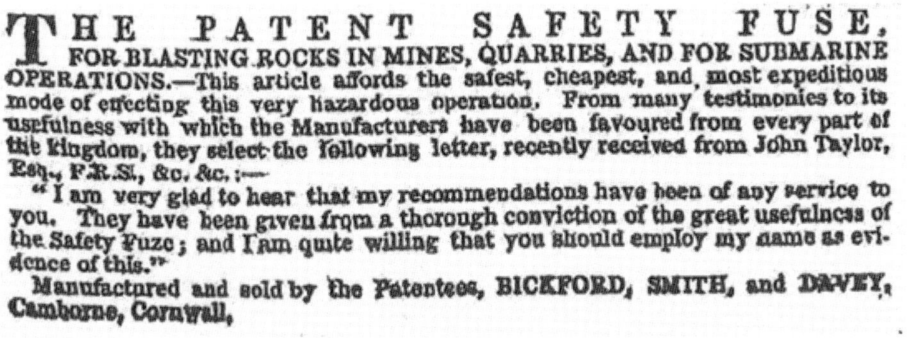

> **THE PATENT SAFETY FUSE, FOR BLASTING ROCKS IN MINES, QUARRIES, AND FOR SUBMARINE OPERATIONS.**—This article affords the safest, cheapest, and most expeditious mode of effecting this very hazardous operation. From many testimonies to its usefulness with which the Manufacturers have been favoured from every part of the kingdom, they select the following letter, recently received from John Taylor, Esq., F.R.S., &c. &c.:—
>
> "I am very glad to hear that my recommendations have been of any service to you. They have been given from a thorough conviction of the great usefulness of the Safety Fuze; and I am quite willing that you should employ my name as evidence of this."
>
> Manufactured and sold by the Patentees, BICKFORD, SMITH, and DAVEY, Camborne, Cornwall.

Figure 15. Advertisement from the *Mining Journal*.

In 1835 Henry English began the publication in London of *The Mining Journal and Commercial Gazette*. Smith quickly began to advertise the fuse, with the same advertisement appearing in every issue for several years.

John Taylor (1779-1863), named in the advertisements, was a mining engineer and mine investor. He owned shares in mines throughout Britain. In 1819 he re-opened the Consolidated Mines in Gwennap. Under Taylor, in 1824 the neighbouring mine, United Mines, were worked with Consolidated, and eventually employed over 3,000 people (which included 597 children). From 1819-1840, nearly 300,000 tons of copper were raised here, which sold for £2,099,485. To have gained John Taylor as such an enthusiastic client provided a great impetus for the fuse factory. In 1835, Taylor gave evidence to the Select Committee of the House of Commons on Accidents in Mines. He testified that the safety fuse was one of the very best inventions for preventing injuries. He explained:- 'with the safety fuse he (the miner) may take two or three feet, and so delay the explosion that he may be in a place of safety before it happens. This is made so cheaply that no difficulty has been found in its introduction; the men are supplied with it, so that it is not worth their while to make the common fuses'.[10]

In 1837 the Hayle Railway opened a line between Hayle and Portreath. By 1838, it had extended to Redruth, with one short branch line opened just west of Tuckingmill to North Roskear to serve the mine, and a second opened to North Wheal Crofty Mine, (then part of East Wheal Crofty), just north east of Tuckingmill. The fuse factory was now at the centre of a thriving tin and copper mining area, and had even started (c.1834) a small gasworks on the corner of what is now Tolgarrick Road and East Hill, Tuckingmill, close to the present day A3047.

George Smith was not only occupied with managing the fuse works, but devoted a large part of his life to writing religious pamphlets and books. He remained a lay Wesleyan preacher, and was deeply involved in religious life in Camborne. He became a trustee of Wesleyan Chapel in Camborne in 1841,[11] and contributed funds to build the Camborne Literary Institute in 1842, overlooking Commercial Square.[12] In 1841 he was living in Chapel Street, in Camborne, where he describes himself as a 'merchant' in the census. Thomas Davey (45) was living in Tuckingmill with his wife Mary and seven children, and described himself in the census as a 'safety fuse maker'. The eldest two boys, Thomas (20) and Simon (18) were employed as clerks in the fuse factory.

At some stage Bickford, Smith and Davey decided to expand into northern France. They probably realised that when their patent expired in 1845, they could expect immediate local competition. It was decided to locate in Rouen. There were substantial coalfields there; in addition it could be reached easily by boat travelling across the Channel. A commercial agent living in France, with the Cornish surname of Trestrail, was instructed to buy a site 80 feet long and forty to fifty feet wide, and construct a building. He was to receive £50 until March 1843, and thereafter £100 per year. Eventually he was paid £60 a year and his employee A. Chanu was paid for eight months work a year. A letter written by the Tuckingmill owners at that time made it clear that the French fuse factory could only sell fuse in France and in French overseas territories. The letter also stated that they intended to manufacture in Belgium and in Saxony, in Germany. At the end of 1842, Simon Davey was sent over from Tuckingmill to become General Manager of the factory. He returned to marry Catherine Blamey in the Chapel of Ease, Pool, at the end of June 1845. Her father was John Blamey, who was a clerk in the Tolgullow Office on the very wealthy Williams Estate, Scorrier. Their son Eugene was born in 1848, and Catherine died the following year. By 1853, the partners in the Rouen Company were J. S. Bickford, George Smith, Thomas Davey, Simon Davey and A. Chanu.[13]

Bickford, Smith and Davey proceeded with their plans to start a factory in Saxony, and signed an agreement with Francis and Ernest Jacobi to build and run a factory at Meissen, with the Jacobis to have a quarter share. Joseph Eales, from their new factory in Connecticut was sent to Meissen in 1845 to manufacture the fuse. Output increased from 64 thousand coils in 1848 to 141 thousand coils in 1851.

In 1842, Bickford, Smith and Davey were forced to petition the Vice Warden of the Stanneries of Cornwall. They were owed £36 pounds and 10 pence by Tregothnan Consolidated Mines, Kea, and were looking for an order to compel the purser, Francis Todd of Pendennis Castle, to settle the debt.[14]

In his autobiography, George Smith devoted just a short paragraph to what was to prove a lucrative expansion into Connecticut, USA. He wrote that a gentleman called Bacon called into the fuse factory in Tuckingmill, around the time of the patent trial

Figure 16. Dr George Smith

in Exeter, and asked if he could be appointed as the sole agent for the USA. This led to the formation of a fuse manufacturing company in Connecticut, namely Bickford, Bacon and Co.

In the mid-1840s, Smith and his wife purchased the old Camborne Workhouse, on the western side of Beacon Hill, less than a mile from Camborne Cross and the new railway station. It was also close to Smith's old boyhood home at Condurrow. They had five children by then (a baby girl had died in 1835) – a seventh (George) was to follow in 1845, and an eighth (Henry) in 1849. They converted this house, which they named 'Trevu', into a dwelling house, and laid out beautiful gardens.

However, the patent obtained by William Bickford in October 1831 expired in October 1845. A Tuckingmill miner called William H. Launder, with unnamed

Figure 17. Camborne Cross, the site of the present-day library, c.1890. J. C. Bennetts photo (RIC).

backers, was quickly put in charge of establishing a fuse factory in Redruth, called the British and Foreign Fuse Company, and advertisements started appearing in newspapers in April 1846. In the same year a group of investors started the Unity Fuse Company in Gwennap.

Further afield, Bickford, Smith and Davey and Co. had to support a safety fuse enterprise in the United States. They no longer had a monopoly in fuse manufacturing, and the company had to develop new strategies and new markets in order to withstand the developing competition.

References, Chapter 3

1. G. L. Wilson. *History of Bickford's Fuse*. In 1962, G. L. Wilson, who had access to the Bickford Smith & Co. files, then the property of Imperial Chemical Industries in Ardeer, Scotland, wrote a short history of the fuse. This was probably undertaken internally for ICI. In the foreword he explained that when ICI closed the Tuckingmill factory in 1961 (which they then owned), some old accounts, notebooks and scrapbooks were found, from which he put together a short account. Wilson wrote that much else was lost during World War II when lofts were cleared, due to a fear of incendiary danger. He added that ICI intended to keep the papers at their Ardeer factory – the site of Alfred Nobel's 1870 explosives company. The ICI works at Ardeer, in North Ayrshire, was the major global supplier of explosives. With the

closure of this factory, it is feared these original Bickford Smith records are now either lost, or may have been shipped to Australia where their whereabouts are unknown. A copy of the original document is in the Peter Bickford-Smith Archives, Cornwall.

2. *Ibid.*
3. The autobiography of George Smith. LL.D. Cornish Studies Library, Redruth
4. *Ibid.*
5. This publication is in the Peter Bickford-Smith Archives.
6. *Royal Cornwall Gazette*, May 25, 1833
7. T. A. Morrison. *Cornwall's Central Mines. The Northern District.* 1810-1895. p 311
8. The autobiography of George Smith. LL.D. Cornish Studies Library, Redruth
9. William Bickford's Will. Cornwall Record Office. Ref. No. AP/B/5893
10. *Mechanics Magazine*, Vol. 24, page 412
11. J. F. Odgers. *Wesley Chapel 1828-1958*. The Camborne Printing and Stationery Co. Ltd. C 1959
12. T. R. Harris. *Dr. George Smith 1800-1868*. Pub. Cornish Methodist Historical Association. No. 13. 1968
13. See reference 1
14. Document in the Cornwall Record Office

Chapter 4

From Tuckingmill to Connecticut — Expansion into the USA

The oldest copper mine in the USA is believed to be the Copper Hill Mine, which is located in Simsbury, Connecticut, some 110 miles south west of Boston, Massachusetts. The mine was re-opened in 1830 under the name of the Phoenix Mining Company, and the new owners appointed Richard Bacon as manager.

Bacon travelled to England in the mid 1830s on mining business and visited the Tuckingmill fuse factory. The importation of Cornish-made fuse into the USA carried an import duty of 25 percent, and it was decided to manufacture fuse in Simsbury. A company was formed under the name of Bacon, Bickford, Eales and Company, under the supervision of Joseph Eales.

In 1839, Bickford, Smith and Davey dispatched a young book-keeper in their employ to Simsbury. His name was Joseph Toy, and he was to remain in Connecticut

Figure 18. Ensign, Bickford & Co. Office, Simsbury, Connecticut (Author's collection).

Figure 19. Souvenir issued by Ensign, Bickford & Co. in 1936.

for the rest of his life. Toy eventually purchased the interest in the business belonging to Eales, and became a junior partner. Richard Bacon left the business in 1851, shortly after a disastrous fire killed eight female workers. The business was renamed Toy, Bickford and Company. Joseph Toy oversaw the adoption of new methods of mechanisation, while concentrating on maintaining a very high quality fuse.

Two of his sons-in-law entered the business – Ralph Hart Ensign and Lemuel Stoughton Ellsworth. Ellsworth was sent to California to establish a new plant there, and in 1887 Toy's step-son, James B. Merritt took over the management. The Californian business was eventually taken over by the Coast Manufacturing and Supply Company, located at Trevarno, California.

In 1887, after the death of Toy, his son-in-law Ralph Hart Ensign became manager, and the company was renamed Ensign, Bickford and Company. In 1907 it merged with the Climax Company (a fuse-making company started by Richard Bacon's two sons) and became the Ensign-Bickford Company. The company diversified, and became a large conglomerate. It is still in private hands, and wholly American owned.[1]

Joseph Toy
Thanks to the internet, it is possible to know some facts about Toy's life in Cornwall.[2] He was born in Roskear in 1808, the son of Robert and Ann Hosking Toy, and the

Figure 20. Joseph Toy.

youngest of eight children. After his father died, he and his mother went to live with his brother John Toy, who became captain of West Seton Mine, Camborne. He married Jane Ostler, a Camborne milliner, in 1833, and three children soon followed.

Bickford, Smith and Davey, who employed Toy, were anxious to have accurate financial statements for the newly established fuse manufacturing concern in Connecticut, and offered Toy a position there as book-keeper. In the summer of 1839 the family travelled, via Falmouth, to Portsmouth and embarked on a six week voyage. From New York they sailed to Connecticut, and travelled the final ten miles to Simsbury on a wagon piled high with luggage. They never returned to Cornwall.

References, Chapter 4

1. The Ensign-Bickford Company, Simsbury, Connecticut. 100 Years, being the story of the safety fuse in America since 1836.
2. Mary Harris Toy Dodge. The story of the Toys. www.gutenberg.org (ebook)

Chapter 5

Bickford, Smith and Davey — the George Smith Years 1846-1868

To the Adventurers and Agents of Cornish Mines, and the Public

Scarcely any part of the practical economy of Mining has been watched over with more careful jealousy than the connection of Mine Agents with the supply of materials in Mines. For many years, therefore, Mine Agents have altogether abstained from the vending of Mining Stores, or they have kept their proceedings relating thereto as secret as possible.

With respect to price …. fuze could be sold at a halfpenny per foot or one shilling per coil of 24 feet. As its sale increased, we reduced the price, first to 9 pence, then to 8 pence; at this price the consumption in Cornwall was as great as it has been since. From 8 pence we brought the price to 6 pence and then to 5 pence, which was the price charged at the expiration of our first patent. No greater proof of the moderation of this price can be requisite than the fact, that our opponents explicitly state, that they "do not compete with us on price".

But it may be said, that the Fuze is now an article of open trade, and must as such be open to competition. A certain number of Mine Agents and others holding places of profits in Mines, are associated together as companies – manufactories formed – the article is made; and each partner without avowing himself, labours insidiously to shut us out of the market, and to introduce into the Mine, which he is paid to conduct, an article, in the sale of which he has a pecuniary interest.

We ask fair play, and have the fullest confidence that Cornishmen will not sanction, in our case, and to our prejudice, a course , which if generally adopted, would be as fatal to the Mining interests, as it must be to the Mercantile interests of the County. *June 1846.*[1]

Bickford, Smith and Davey were now open to the full forces of competition. They suddenly found this difficult to accept, and sent a printed letter to influential customers. They blamed certain mine agents for not letting their fuse into the mines which these particular agents managed. Competitors had started the British and

Foreign Fuse factory in Redruth, in the Plain-an-Gwarry area, and an even more serious threat had emerged around 1847 in the St Day area, in the shape of Hawke's Unity Fuse Company. In the same year Bruntons began manufacturing fuses in Penhellick, near Pool. Later still, William Bennett was to establish a rival factory almost within sight of Bickford and Co. It was probably to counteract these very real threats that young Simon Davey was sent to Rouen, in northern France, to establish a fuse factory to serve the coalfields there.

It was during this period that the Tuckingmill Foundry began to expand. In 1841, Edward Vivian was listed as a 'fuse factor' or broker.[2] He was the son of William Vivian, the Tuckingmill factory founder. John Solomon Bickford, the co-owner of the fuse works, had moved from Hayle in 1838 with his wife Sarah (Davey) and two children and taken up residence in a large house in Tuckingmill, together with two female servants.[3] He described himself in 1851 as a 'Safety fuze manufacturer employing 25 persons'.[4] In 1859 George Smith was appointed Chairman of the Cornwall Railway, and on May 2nd led a delegation from West Cornwall to the opening of the Royal Albert Bridge at Saltash, where he met Prince Albert.

> PATENT SAFETY FUSE.—The GREAT EXHIBITION PRIZE MEDAL was AWARDED to the MANUFACTURERS of the ORIGINAL SAFETY FUSE, BICKFORD, SMITH, and DAVEY, who beg to inform Merchants, Mine Agents, Railway Contractors, and all persons engaged in Blasting Operations, that, for the purpose of protecting the public in the use of a genuine article, the PATENT SAFETY FUSE *has now a thread wrought into its centre*, which, being patent right, infallibly *distinguishes it from all imitations*, and ensures the continuity of the gunpowder.
> This Fuse is protected by a Second Patent, is manufactured by greatly improved machinery, and may be had of any length and size, and adapted to every climate.
> Address,—BICKFORD, SMITH, and DAVEY, Tuckingmill, Cornwall.

Figure 21. Advertisement in the *Mining Journal*, December 3rd 1853

In 1846, two patents were taken out, which described spinning three fuses at once, using a centre thread, and a white varnish and gutta-percha coating.[5]

In the same year, on the 16th of February, a new deed of business was signed. The business was divided between John Solomon Bickford, George Smith and Thomas Davey, with the latter to pay the difference between a ¼ and a ⅓ of the valuation. This agreement was to run for 20 years, but a new 14 year agreement was signed in 1853, admitting Francis Pryor "in consequence of payment and mutual trust". Davey and Pryor were sleeping partners, and the firm's name was changed to 'Bickford, Smith, Davey and Pryor'. In the terms of the agreement, John Solomon Bickford and George Smith were to receive a management salary of £150 per year, and the rest was to be divided into ¼ shares.[6] Francis Pryor was a mine proprietor, mine agent and purser of Redruth.[7] At the end of the 14 years, Pryor's association with the firm ended. Why he was taken on as a partner is not known, but it could have been done to procure funds from him in order to invest in the Connecticut factory.

The firm decided to enter a sample of their fuses into the Great Exhibition in London, in 1851.[8] This was held in the newly built Crystal Palace in Hyde Park.

Entry 424
BICKFORD, SMITH & DAVEY, Tuckingmill, Cornwall,
Inv. and Manu. – Several kinds of safety fuze, adapted to convey fire to the charge in blasting rocks or ice, or in submarine operations.

They were very successful and came away with a medal, which was publicised widely.

The red thread which was now running through the centre of their fuses was to become a part of their trademark. A second thread was added later, in 1865.[9]

In 1851 Thomas Davey senior was living in 'Tolgarrack' with his wife Mary (Thomas) and two daughters. His youngest daughter Emma died in September 1852 at Penlu, Tuckingmill, aged just 19 years. Thomas Davey died just four months later, on December 31st 1852, aged just 59 years. The *West Briton* newspaper described him as 'a leading and useful member of the Wesleyan Society'.[10]

His son Simon Davey at this time was in Rouen in Northern France, where he had established a fuse factory, and had settled down with his second wife Jeanne Emma Augustine Sarrasin de Maraise. Thomas Davey junior was living in Tuckingmill, where he was the proprietor of a rope making factory, employing 19 persons. His brother Charles Davey, then aged 16, was attending a residential school in Falmouth.[11] It seems that Thomas Davey junior, then aged 31, assumed his father's role in Bickford, Smith and Davey, although he kept his rope making business in Tuckingmill.[12]

He also had a share in another business. The concern, known as the 'Phoenix Ropewalk', Illogan, appeared in *Perry's Bankrupt and Insolvent Gazette* on January 25th, 1848. The owners were listed as John Solomon Bickford, George Smith, Thomas Davey and Elias Christian. The ropewalk was situated east of Tuckingmill, with the branch line of the old Hayle to Portreath railway running alongside.

There was better news for George Smith in April 1852, when his eldest son William Bickford Smith married Margaret Leaman Venning in Broadhempston, Devon. William was the proprietor of the Camborne Gas Works, and he and his new bride took up residence in Redbrook House, on Beacon Hill, close to his parents.[13] In 1862 he and William Bennett, an engineer, took out a patent. It was for the invention of improvements in the method of, and apparatus for, preventing the injurious effects occasioned by smoke, sulphur and the deleterious gases which escape from stacks, chimneys etc.[14] It seemed the residents of Camborne may have been complaining about the 'rotten eggs' smell from the gasworks. Bennett was later to become a serious rival in the fuse manufacturing business.

Tragedy struck in June 1853, when John Solomon Bickford's son John died, aged just 18 years.[15]

Simon Davey, although believed to be living in Rouen, France at that time, took out a patent in 1858 for 'an improvement in the manufacture of safety fuzes for mining and military purposes', where he described himself as a 'safety fuse manufacturer, Illogan.[16]

In early November 1858, Bickford's fuse received widespread publicity when it was used to blow up the hazardous Vanguard Rock at the entrance to Devonport harbour, near Plymouth. A large metal cylinder, fourteen feet in length and four and a half feet in diameter, containing a ton of gunpowder, was placed underwater in a cavity in the rock. This cylinder had been shipped out to the Crimea around 1854, to be used in the blowing up of the fleet at Sebastopol, but was shipped back to England unused. At the third attempt, thirteen minutes after Bickford's fuse was ignited, the cylinder exploded, sending up a cone of water forty feet high and a hundred feet in diameter.

A tremor resembling an earthquake was felt all along the shore of the harbour, and hundreds of fish were killed. Thousands of spectators apparently cheered after the explosion. A sketch by Major Bredin of the Royal Artillery appeared in the *Illustrated London News* shortly afterwards.

The fuse factory was now a large employer of female labour. In the 1861

Figure 23. Blowing up the Vanguard Rock, Plymouth

Figure 22. Bickford's trademark.

census John Solomon Bickford described himself as a manufacturer of safety fuse, employing 66 women. [17]

In 1861, a competitor patented a waterproof tube made from lead, or any soft metal alloy. The *London Gazette* of May 12th 1865 carried details of this invention, credited to Joseph Victor of Wadebridge, a hydraulic engineer, James Polglase of Bodmin, a mine agent and William Rounsevell of St. Breock, a carpenter. However, the Office of the Commissioners of Patents for Invention published details of the patent (dated March 22nd 1861), where it is credited only to Polglase and Victor. It was described as follows:- 'the combustible material is enclosed in a small tube of lead or soft alloy, which tube is drawn out between rollers, by means of which the combustible material is compressed and its slower combustion assured. Lengths of this tube may be cut off as required'.

In the 1861 census, James Polglase was described as a 'metallurgist and mineralogist'. Manufacture of the fuse was started in Wadebridge (on the site of what is now the Irons Brothers Foundry). They entered their 'metallic safety fuses' into the Great Exhibition of 1862 in London, stating in the catalogue that the fuses were tested in blasting operations before the Miners' Association of Cornwall and Devon at the Royal Polytechnic Exhibition, and obtained the Society's prize medal. It appears that the business was sold shortly afterwards for £4,000, and stopped production. It may have been sold to a rival firm, which bought out the patent, and closed the fuse factory.

Thomas Davey's Gunpowder Works at Nancekuke

In July 1859, Thomas Davey junior applied to the courts for a licence for a gunpowder mill and magazine at West Wheal Towan, Nancekuke, west of Porthtowan. Davey proposed to use the Accounting House and adjoining buildings of a tin mine that had closed about two years earlier. He told the court that the site 'was within a few yards of the edge of the cliff overlooking the Atlantic, and was on waste common, with no house near it'. Davey, together with his brother Simon, intended to manufacture blasting powder by a new process which he had patented.[18]

Unfortunately, on Tuesday September 10th 1862, an explosion in the drying house at the Powder Works killed six women workers. Davey thought that the gunpowder in the front of the house was ignited by lightning through the partially open windows.[19]

The Gunpowder Mill was closed. Ernest Landry, who was born in 1889 in a nearby farmhouse known as 'Factory Farm', explained that the Basset's Tehidy Estate later had all of the factory buildings demolished.[20] They were then rebuilt further inland into this farmhouse and farm buildings in the shape of an 'L'. There is now an arable field where the Gunpowder Mill once stood.

> **Davey's Patent Blasting Powder.**
> *Manufactured by DAVEY BROTHERS, and Co., Nancekuke Powder Works, Tuckingmill, Cornwall.*
> THIS BLASTING POWDER has the following advantages over every other in use.
> Its *combustion* is *slower* and more *perfect* when confined in the hole.
> It is more impervious to moisture.
> Produces less smoke. Is less dangerous.
> It bursts as *much rock*, with a *charge* occupying the same, or even *less space* ; and its *weight* being 20 to 25 per cent. *less* than ordinary gunpowder, a saving of *one-fourth* of the *cost is* effected.
> Davey Brothers, and Co. beg to state that this powder is specially made for blasting, and, from its slow combustion, is not adapted for projectiles. They would therefore caution consumers not to be induced by interested parties to put it to a fallacious trial, by firing a ball from a mortar, which is no test of its explosive force when confined.

Figure 24. The *West Briton*, June 6 1862

In February 1863 Thomas Davey junior died aged 42. In the 1861 census he had described himself not only as a powder manufacturer employing 10 persons, but also as a rope maker employing 17 persons. Simon Davey, his brother, was then brought into the management of Bickford Smith, although he was living in Rouen in northern France. Thomas Davey's wife, Anna, who had been left with six young children including a 9 months old baby girl, put their elegant house in Penlu, Tuckingmill, up for sale. The auctioneer was H. V. Newton, Camborne. At the time she was living next door to John Solomon Bickford.

It is interesting to read the sales notice.[21] These houses had been built to house the local business owners sometime in the early 1850s, and were very luxurious.

For Sale: - PENLU VILLA, containing spacious lobby, large drawing room

Figure 25. Penlu villas, c 1895. North Roskear mine is in the background. (RIC)

and dining room, breakfast room, sitting room, two kitchens, pantry and dairy on the first floor, six excellent bedrooms and water closet on the second floor, together with necessary outbuildings, large walled garden, well-stocked with fruit trees etc., and green-house and hot-house in same, furnished with well-fruiting vines, very neatly arranged and well grown shrubberies, ornamented with a fountain, grotto etc., together with an excellent stable yard etc.; also a good meadow, about half an acre adjoining.

Anna Davey then went to live with her sister and brother-in-law, a draper called Robert Parkyn, in Kenwyn, near Truro.[22]

John Solomon Bickford, whose first wife Sarah died in 1861, remarried in July 1866 at the age of 62. His bride was Julia Vivian (40), the daughter of mine agent Henry Andrew Vivian. He was the son of Andrew Vivian, Richard Trevithick's partner in driving the famous road locomotive up Camborne Hill on Christmas Eve 1801.

In November 1866 the partnership with Francis Pryor was dissolved. The notice, which was inserted into the *London Gazette* of November 20th, 1866 stated that the business was owned by John Solomon Bickford, George Smith, William Bickford Smith, and Simon Davey, 'by whom all debts due to and owing by the said firm will be received and paid'. The firm's name had changed also, to Bickford, Smith, and Co.[23]

On Saturday 28th August 1868, George Smith died at his Trevu residence. The details of his will were published in *The Royal Cornwall Gazette*. He was described as 'a safety fuse and arsenic manufacturer'. The report stated that his personalty (i.e. personal estate) was valued at less than £50,000, not including the freehold property. The newspaper reported that his interest in the fuse company passed to his sons.[24]

George John Smith (23) now joined his older brother William Bickford Smith in the management of the Tuckingmill fuse factory.

References, Chapter 5

1. Letter to adventurers, June 1846. Cornish Studies Library, Redruth

2. Edward Vivian. Census 1851
3. *The West Briton and Royal Cornwall Advertiser*, Thursday evening. February 24, 1870. Obituary
4. John S. Bickford. Census 1851
5. See reference 1, Chapter 3. G. L. Wilson
6. Ibid.
7. Francis Pryor (1819-1870) was born at Bolitho Farm, Crowan, the son of Captain William Pryor. At one time he had the management of fourteen mines, including Tincroft, West Caradon and St. Day United. His obituary in the *West Briton* (22.12.1870) stated that his large connection with mining was the means of his becoming a partner in Bickford, Smith and Co., which he continued 'for many years'.
8. Official Catalogue of the Great Exhibition 1851. Exhibit 424.
9. *Royal Cornwall Gazette.* July 21, 1882
10. *The West Briton and Royal Cornwall Advertiser.* January 7, 1853
11. Charles Davey. Census 1851
12. Thomas Davey junior. Census 1851
13. Camborne Post Office Directory, 1856
14. *The London Gazette*, May 9 1862
15. *The West Briton and Royal Cornwall Advertiser*, June 17, 1853
16. *The London Gazette*, July 23, 1858.
17. The numeral and word '66 women' are written on the original census returns. These can be seen on www.ancestry.co.uk
18. *Royal Cornwall Gazette*, July 8, 1859
19. *The West Briton*, September 12, 1862
20. Landry, Ernest. *Memories of Nancekuke*. 1978
21. *Woolmer's Exeter and Plymouth Gazette*, May 15, 1863
22. Census 1871
23. *The London Gazette*, November 20, 1866
24. *Royal Cornwall Gazette*, October 29, 1868

Figure 26. The Smith family, probably at the wedding of William Bickford Smith (centre). 1852 PB-S Archives

Figure 27. Cart entering Bickford Smith's factory c 1902

Figure 28. Tram passing Bickford Smith's factory c 1905. The single storey building on the left was removed c 1912.

Figure 29. Red River Valley c 1905. Bickford Smith's fuse factory is centre left. RIC

Chapter 6

Bickford, Smith and Co. — The Next Generation Takes Charge and a Terrible Accident Occurs. 1846-1868

In 1869 John Solomon Bickford decided to enlarge his house at Penlu, overlooking his fuse factory. At the end of November, he celebrated the completion of the walls by giving a dinner and tea to over forty of the labourers. Mrs Sleeman, who managed the Tuckingmill Hotel, provided the meal.[1] The hotel had been built in Tuckingmill in1850. The first lease had been signed with the owners of Redruth Brewery in September 1850.[2]

If he had enlarged his house to accommodate the baby his wife was expecting, then, sadly, he did not live to see his new son. John Solomon Vivian Bickford was born on the 26th of April, 1870, two months after his father's death.

John Solomon Bickford died on the 17th of February, 1870. He was the only son of William Bickford, the inventor of the safety fuse. He had been unwell for two weeks, and had complained of head pains. Two days before his death he had attended a meeting of Wheal Margaret adventurers in Tabb's Hotel, Redruth, and a day later attended a meeting of Nangiles Mine investors. It was at this meeting that he complained of 'deadness in his arm and cold in his body'. On returning to his home, he died suddenly.

His obituary[3] reported that he devoted himself, until 1859, to the development of the fuse works, and had travelled twice to Europe and once to the USA. He apparently was a successful investor in mines.

In 1859 he was chosen as Captain of the newly formed Rifle Volunteers, Camborne.[4] At a cost of £3,000, he paid for an armoury, an armourer's house, a gymnasium and a drill ground at North Roskear. The *West Briton* reported that he attended the Corps for five nights a week until his hearing began to fail in 1867. He was made a Major of the Corps, and was known from then on as 'Major Bickford'.

John Solomon Bickford was buried in the cemetery of All Saints Church, Tuckingmill, close to the grave of his first wife Sarah and son John. The entire Camborne Rifle Corps, in full uniform, marched four abreast ahead of the funeral procession, led by Captain Pike. Following the funeral cortege were the fuse factory employees, and the workmen of the nearby Tuckingmill Foundry, in which he was a partner.

Twelve months after John Solomon Bickford's death in February 1870, the *Royal Cornwall Gazette* carried a report of the sale of his properties. He died intestate. At the start of February, his properties were sold at Abraham's Hotel (now Tyack's) Camborne. Lot 1 was a moiety (a half) of the hotel. His residence on Penlu Terrace (described as 'a mansion') found no bidders. Two dwelling houses on Penlu Terrace (numbers 4 and 5), with stables and 'smithery' were sold to William Bickford Smith (his nephew) for £350, who also bought a moiety in Camborne Gas Yard for £76. Number two was bought by a Mr. Willoughby for £315, while number 3 was bought by the vendors. The house and drill ground of the 2nd Devon and Cornwall Rifles were sold to George John Smith (another nephew), who stated that he wanted the volunteers to be able to use it as they had done when John Solomon Bickford was alive. The following year his widow Julia returned to live near her family in St. Hilary, Cornwall.[5]

Figure 30. John Solomon Bickford. PB-S Archive.

The management of the Tuckingmill factory had now passed completely to the grandsons of William Bickford, namely William Bickford Smith and George Smith. In the 1871 census, William Bickford Smith (42) was living in Redbrook House on Beacon Hill with his second wife Anna, where he is described as:- 'J.P., Capt. Devon and Cornwall Rifle Volunteers, Landowner and patent Safety Fuse manufacturer'. He was clearly the head of the business, as a new patent in June 1870, was registered in his name. It was for an improved coating for the safety fuse.[6]

Henry A. Smith (23) was still living with his widowed mother Elizabeth Burall Smith (65) in Trevu, also on Beacon Hill, and described as a 'Flax merchandiser',[7] while his brother George John Smith (25) lived next door, described as a 'manufacturer of Safety Fuse'. It seems that Henry had taken over the management of the jute spinning factory in Pengegon started around 1866 by his father George, and which at that time could process 7 hundredweight of jute a day.[8]

The Easter Saturday Tragedy, 30th March 1872

The Royal Cornwall Gazette of Saturday morning April 6th, 1872 carried the following report:-

FEARFUL CALAMITY AT TUCKINGMILL
EIGHT GIRLS SUFFOCATED

The safety fuse manufactory of Messrs. Bickford Smith, and Co., at Tuckingmill, a suburb of Camborne, employs about a hundred young women and girls, for whom the work of spinning the tube which encircles a small quantity of gunpowder is well adapted. Wherever this explosive compound is extensively used, a certain amount of danger there must be; but the late Major Bickford and Dr Smith gave years of thought and attention to precautions, as to buildings, machinery, and the rules which govern the conduct of the workpeople themselves, so as to reduce the hazard to a minimum. That these safeguards were excellent is proved by the immunity of the Tuckingmill factory for a period of 38 years from serious disaster. And since the property has passed to Capt. Bickford-Smith and Mr. George Smith no regulation has been relaxed whereby such a sad result as that we are about to describe could come about.

The employees had a holiday on Good Friday, and on Saturday word was given that the engines had ceased work, in order that the boilers might be cleansed. The women, therefore, could 'clean up' their respective rooms and machinery, which really meant careful sweeping of every crevice of the apartments, polishing the machinery entrusted to them, and a general tidying of surroundings, so as to commence labour in earnest on Monday. The apartments, of which there are four or five under one roof, consist of a reeling mill on the ground floor and spinning mills on the first floor. In the latter small quantities, fine as a thread, are run into machinery, and out between the revolving threads, so as to form a small coil. In the afternoon all unused gunpowder is carefully collected and returned to one of the overseers, who sees the ingredient conveyed to an isolated machine. The apartment has four pairs of spinning apparatus, and is well lighted by three windows on each side. It is about 25 feet long and 22 wide, and there is ample room to move all round the apartment, between the machinery and the walls. The staircase-head comes up in one corner of the room and lessens this space. At the opposite corner were temporarily placed some 20 coils of fuse, each about 24 feet long, and containing about 3 pounds of powder. These caused the fatal results of the disaster.

Sixteen girls – whose outer attire and shoes are taken off in rooms specially devoted to that purpose, a little distance from the scene of their labours – were engaged in cleaning the machinery, two others had come up to remove the coils of fuse, and another was there to see a friend, making 19 girls in the room, most of them being at the lower end. To pass from this spot (round one side of the room, between the machinery and the wall) they would come to the corner where the safety fuse was placed, and, passing this, would gain the door in the next corner. The windows could be pushed open easily, as they pivot at the top, and a jump of 10 or 12 feet would land a person in the yard. What, however, could be seen at a glance in cool moments afterwards, seems never to have occurred to the bewildered minds of the poor girls, who, at one moment are recounting, with laugh and joke, the incidents of the Good Friday holiday, the next are a panic stricken herd, and the next are senseless and dying.

What caused the accident was not at first known. It was surmised that, in the act of sweeping, a spark was struck, fell into a crevice in the floor, and thus ignited a very small train of gunpowder, and so led to the firing of the safety fuse. Suddenly there was a blaze, then a rapidly accumulating and dense volume of smoke, and then the stampede for life. Those nearest the door escaped, the very first of the more fortunate frightened only. The next batch, less nimble or not so well situated for a rush, were burnt as they passed the fuse in the corner. In this way eight or ten made good their exit. At last – either stumbling near the fuse, or faint with excitement, or overcome by the fumes – one fell. Two or three met the same fate. One girl, with more presence of mind than her companions, at the first flash of fire rushed to the staircase, dashed down the partition, which led to the stairs, and was saved. In all, eight young women perished from suffocation. Some of the bodies were untouched by fire – some much scorched; some faces looked calm and ruddy, hours after death – others were burnt, and seemed as if the hapless owners had died in agonies.

The alarm having been given, and smoke been seen to pour through the windows, a hose was quickly attached to the Camborne water-pipes and a fine stream of water directed on the building. Previously to this, however, and over and over again, the inquiry was made – Are all the girls safe? The answer was "Yes". Then someone discovered that a spinner was missing, but almost immediately she presented herself. The escaped ones said they were all out; and so efforts were made – not to risk entering the room as many present would have unhesitatingly have done, but to check the fire. As the smoke continued to pour out in volumes, the windows and roof were broken, so that the water might have more effect. Some fifteen or twenty minutes passed in this way – the men at the roof and windows having no idea of the dread sight to be presented too quickly. Suddenly it was discovered that some young women were missing. Instantly the room was entered, and one after another of the sad heap was dragged out – in all eight lifeless corpses, and one who happily recovered. Medical aid was at hand, but was of no avail whatever except for those who rushed out so promptly. These are being cared for by Captain Bickford-Smith (sic) and Mr. George Smith in the kindest way – indeed they have personally seen that everything shall be provided for them. The actual damage to the building and the machinery is nothing beyond that caused by the water, the fire having only slightly charred one board beneath the coils of fuse. But the gloom which has fallen on relatives, friends, and employers, saddens the whole neighbourhood.

The Inquest
The Royal Cornwall Gazette, of Saturday April 6th 1872, carried a full report of the inquest, which took place in Abraham's Hotel (now Tyack's) in Camborne on Tuesday 3rd April:-

THE MELANCHOLY ACCIDENT AT TUCKINGMILL

On Tuesday the inquest on the bodies of Emily Clemo, 22, Ellen Goldsworthy, 19, Mary Ellen Sims, 18, Louisa Ann Sims (her sister), 16, Emily Carah, 21, Elizabeth Ann Marks, 17, Annie James, 17 and Martha Towan, 17, was held before Mr. G. P. Grenfell, deputy-coroner. Mr J. R. Daniell, solicitor of Camborne, appeared to watch the proceedings on behalf of the owners of the factory, and Mr. Bickford-Smith and Mr. George Smith were also present.

Evidence was given firstly by Samuel Hocking, the engineer in the fuse factory. He described the scene. He stated that no work had been done the previous day, Good Friday, and that at 7 o'clock on Saturday morning the fireman commenced cleaning out the boiler. He himself had been in the passage of the boiler room, close to the spinning room where the explosion had taken place. He heard a little puff or slight explosion. He stated that there had been no fire in the factory for the previous forty hours. He described the attempts to put the fire out in the No. 2 spinning room, and heard several girls say that they had all escaped.

Emily Richards, an overlooker, then gave evidence. She said she looked after about 40 girls in the two spinning rooms and reeling-rooms. She said there should have been eighteen girls cleaning that morning, eight from the spinning room and eight from the room below, together with two girls who were there to remove the reels of fuse. She said that every bit of powder had been taken out on Thursday night – "it is taken out every night and the floor thoroughly swept". She explained that on the Saturday of the explosion the coils of fuse which had been spun on the Thursday afternoon were still in the corner of the No. 2 spinning room, where they were placed last thing on Thursday night.

She described how the machinery which the girls were cleaning ran down the centre of the room. "The top and bottom are taken off the machines in cleaning;

the screws are sometimes turned with finger and thumb, and sometimes with a spanner".

Another overlooker, Annie Trevarton, then gave her evidence, saying "I heard a noise of something falling ……..the fall I heard was like a piece of iron falling against another piece of iron. Instantly after this the smoke issued from the window".

Emma Miners then gave her evidence: - "I work in the powder-house; that is where the powder is dried before it is taken to the spinning rooms. I was in No. 2 when the accident occurred, but I had no other business there. I was not assisting them, but merely chatting. I heard something fall and immediately saw a spark of fire, which flickered about the room from one place to another".

Another factory girl, Martha Plint, was then called, saying she was in No. 2 room helping to clean the machinery. She described how she heard something fall. She explained that the girls, for the purpose of cleaning, unscrewed the 'slides', which were screwed on by a nut, and which were made of iron.

The engineer, Samuel Hocking was recalled, and stated that he had found afterwards a guide (slide) from one of the machines on the floor. The slide was $\frac{5}{8}$ths of an inch in diameter and 3 feet long. (1½ centimetres and about a metre) He said that in many years' experience he never knew iron falling against iron to produce a spark. "I have not made up my mind as to what the cause of the fire was". He further told the Coroner that it is not the duty of the girls to unscrew the guides. They have no business to do so; if this is to be done a mechanic would be sent to do it.

A crucial witness was then called, named as Sarah Ann Cock, who was working in No. 2 room at the time of the accident. "In the course of cleaning we unscrew the 'arms'. I did not unscrew it. Ellen Sims, one of the deceased girls, unscrewed one of the arms. I saw her do it, and when she had unscrewed it, it fell out of her hand upon some powder dust on the floor. There was scarcely any powder dust there, for I assisted in brushing up the floor on Thursday. I saw the iron arm fall to the ground, and there was at once a flash of fire from the place. There was a slight explosion"

Mr. George Smith explained that the 'arm' was the same as the 'guide' and the 'slide'.

Charles May reported that he had heard a groan, and crept on his hands and knees to the top of the stairs. There he found two girls one on top of the other; when dragged out they were both alive but the uppermost girl died shortly afterwards.

Mr. Thomas Hutchinson, a surgeon, stated that the cause of death was suffocation, caused by the smoke from the burning fuse.

The coroner said he thought they had now got all the evidence they required. He suggested a verdict, which was accepted after some remarks from the jury members, which were all highly complementary to the proprietors for the great caution with which the factory was managed, and for their uniform kindness towards their workpeople. The following verdict was unanimously agreed:-

That the deceased were accidentally suffocated and stifled by the smoke arising from a fire at Messrs. Bickford Smith and Co.'s. Patent Safety Fuse Factory; and which fire was caused by a piece of iron accidentally falling from the hand of Ellen Simms and coming into contact with some powder dust lying on the floor of the building, and which powder dust ignited some fuses placed in the corner of the room; and the jury wish also to express their decided opinion that no blame is due to the proprietors of the said Patent Safety Fuse Works.

The newspaper named the injured girls as:- Elizabeth Ivey; Elizabeth Jane Vincent, Illogan; Mary Ann Matthews, Pool; Sarah Ann Cock, Tuckingmill; and Fanny Bennetts, Camborne. (The 1871 census shows that Elizabeth Ivey (25) lived in Moor Street, Camborne, next door to Emily Caral, one of the deceased girls.)

Gunpowder

What became clear from the evidence given at the inquest were the references to 'powder' or 'powder dust'. What was referred to almost casually was, in fact, the gunpowder used to trickle down the centre of the spinning fuses. Gunpowder or 'black powder' (to distinguish it from Alfred Nobel's patented explosive in 1867 known as 'dynamite') is a mixture of saltpeter (which releases oxygen when exposed to heat) and the fuels needed to burn, namely sulphur and charcoal. When ignited, gunpowder's stored energy becomes the thermal energy of fire and the physical energy of compressed gases. If that energy is enclosed (as in the bored holes drilled by the underground tin and copper miners), with a fuse inserted, and then lit, then the gas will generate such enormous pressure that the rock walls are blasted apart. However, when ignited unenclosed, (such as the loose gunpowder which was lying on the fuse factory floor), the gunpowder burns with a soft thump (possibly the 'puff' which Samuel Hocking heard) and an eruption of flame, and emits a cloud of thick white smoke. It was this smoke, and the smoke from the stored fuses, which also ignited, which suffocated the eight young women. The gunpowder in question would have been transported to Bickford Smith's in wooden barrels from the gunpowder works in Kennall Vale, near Ponsanooth.

The Funeral

The *Royal Cornwall Gazette* described the funeral of six of the women on the Tuesday afternoon of the inquest. Four were buried in the parish churchyard, and two in the Centenary Chapel burial ground. The remaining two women were laid to rest the

	Age	Job in factory	Address	Father	Mother	Sisters	Brothers
Ellen SIMS	17	Spinner	Beacon Near school		Mary (47) (widow) dressmaker	Susan Cock (23) Sarah (12) Carrie (11) Julia (10) Edith (7)	William (19) Henry (12) Both miners
Louisa SIMS (sister of Ellen, above)	14	Reeler					
Emily CLIMO Or Clemo	24	Spinner	Trelowarren St, Camborne	Richard (62) tin miner	Prudence 49		F. S. (11) Tin miner
Emily CARAL (also spelled Carah)	23	Reeler	Moor St, Camborne	William (47) Tin miner	Elizabeth (46)	Elizabeth (15) Working fuse factory	William (19) Henry (12) Both miners
Eliza Ann MARKS	17	Reeler	East Charles St, Camborne	Thomas (50) Engine driver	Elizabeth (54)	Ellen (15) Tailoress	Benjamin ( ) Mine labr at surface
Annie JAMES	18	Reeler	Centenary Row, Camborne	Father abroad	Ann (46)	Grace (22) Dressmaker Eliz (20) Milliner	
Ellen GOLDSWORTHY		Spinner	Sea View Terrace, Beacon Hill		Elizabeth (51) (widow)		Thomas (14) Surface lab mine Joseph (20) Miner
Martha TOWAN	16	Reeler	Trelowarren Street	Father abroad	Mary (59)	Sarah Ann, (23) Eliza (19) Both worked in fuse factory	

Women killed in 1872 explosion at Bickford Smith and Co.'s Fuse Factory, Tuckingmill.[9]

following day, one in the churchyard and the other in the Centenary burial ground, Camborne. 'Immense crowds of people attended the funerals and the greatest of sympathy was shown for the unfortunate girls and their afflicted families. The great kindness displayed by Mr. Bickford Smith and Mr George Smith throughout the whole of the distressing affair is very generally spoken of by the working people with affectionate warmth.'

References, Chapter 6.

1. *The Royal Cornwall Gazette*, September 9th 1871. The 'To be Let' notice stated that the Hotel was 'now in the occupation of Messrs. Magor, Davey and Co. and their under-tenants, whose term of 21 years expires at Michaelmas next.' Magor and Davey owned Redruth Brewery. The notice was dated 'Pendarves, 21 August, 1871.'
2. *The Royal Cornwall Gazette*, November 27, 1869
3. *The West Briton*, February 24, 1870
4. The Rifle Volunteer Corps were created throughout England, Scotland and Wales in 1859, as a result of the formation of the Volunteer Force. The 2nd Corps (1859-80) was formed in Camborne, with John Solomon Bickford as captain, with Thomas Hutchinson as lieutenant and Walter Pike (son of Robert Hart Pike, an influential mine purser) as ensign. The first son of Dr George Smith, William Bickford Smith, was also a Captain. Westlake, Ray. *The Rifle Volunteers*. Pen and Sword Books Ltd. ISBN 978 1 84884 211 3
5. Cornwall census 1871. G. L. Wilson (see reference 1, chapter1) wrote that Julia Bickford was paid £6,348 for a quarter share in the French company in Rouen by William Bickford Smith, and £1,350 for a quarter share of the company Bickford's had started in Germany.
6. *The London Gazette*, .June 27, 1873
7. The Post Office Directory of 1873, under the heading 'JUTE MILLS' lists Smith H. A. & Co., Pengegon.
8. See Reference 1, Chapter 3. G. L. Wilson
9. Cornwall census 1871

CHAPTER 7

Bickford, Smith and Co. - The Years of Expansion
1873-1888

The 1873 Post Office Directory listed not only Bickford, Smith and Co., but also their competitors.

> William Bennett, Tuckingmill
> British and Foreign, Redruth
> Wm. Brunton and Co., Pool
> The Unity Fuse Co., Scorrier

Kelly's Directory of 1873 also listed a subsidiary in Tuckingmill by the name of Bickford, Venning and Co., manufacturers of patent waterproof blasting cartridges, with the Secretary named as William Bailey. Bailey was also listed as the Secretary and Manager of the Camborne Gas Company.

Tuckingmill was expanding – the population in 1871 was 4,108. A school had been built (1843-45), with school teachers in 1873 named as Matthew Bennetts and Miss A. Bawden. Small shops had sprung up – a butchers, a shoemaker, grocers, beer retailers, watchmakers, drapers, and a post office. The elegant Roskear Villas, on the road into Camborne from Roskear Church, had been built by the late 1860s, and were occupied by engineers and managers.

The Coal Mines Regulation Act of 1872, which came into force on 1st January 1873, prohibited miners from bringing loose gunpowder into coal mines. The use of gunpowder-filled cartridges came into widespread use. This lead to the establishment of blasting cartridges being made in registered premises, and seven or eight factories were rapidly put into production. Previous to that, the miners would make cartridges themselves from any available paper. A length of fuse was inserted into the powder at the open end and tied with string. The grease used was most likely that used to grease the pumping engine bearings. In the case of coal mines, where only permitted explosives could be brought underground, the paper was paraffin waxed and the cartridges were then dipped, after charging and sealing, into a container of melted paraffin wax.[1]

In 1873, Bickford, Smith and Co. purchased the powder works and safety fuse business of Charles Davey in St. Helen's Junction, Lancashire.[2] He was the son of Thomas Davey, who helped develop the first fuse with William Bickford. The business in St. Helen's Junction survived for over forty years, supplying collieries with safety lamps and fuses.

Also in 1873, a writer in *The West Briton* newspaper noted that Mr. Llewellyn Newton, aged 89, who was the accountant or book-keeper with Bickford, Smith and Co., walked every day from his home in Camborne to the fuse works in Tuckingmill, where he stood at his desk all day before walking home again in the evening to his lodgings in Basset Street.[3]

In 1874 William Bickford Smith purchased Trevarno, a beautiful estate near Helston. According to a letter in the *Royal Cornwall Gazette*, he paid £25,100 for the house and 819 freehold acres.[4] At the time, he was a magistrate for Cornwall. The old mansion which had formerly been at Trevarno was home to one of the Arundells, the land having been previously owned by Lord Arundell. The new mansion (built c 1834) was rumoured to have cost £7,000. The original owner, G. W. Popham, who

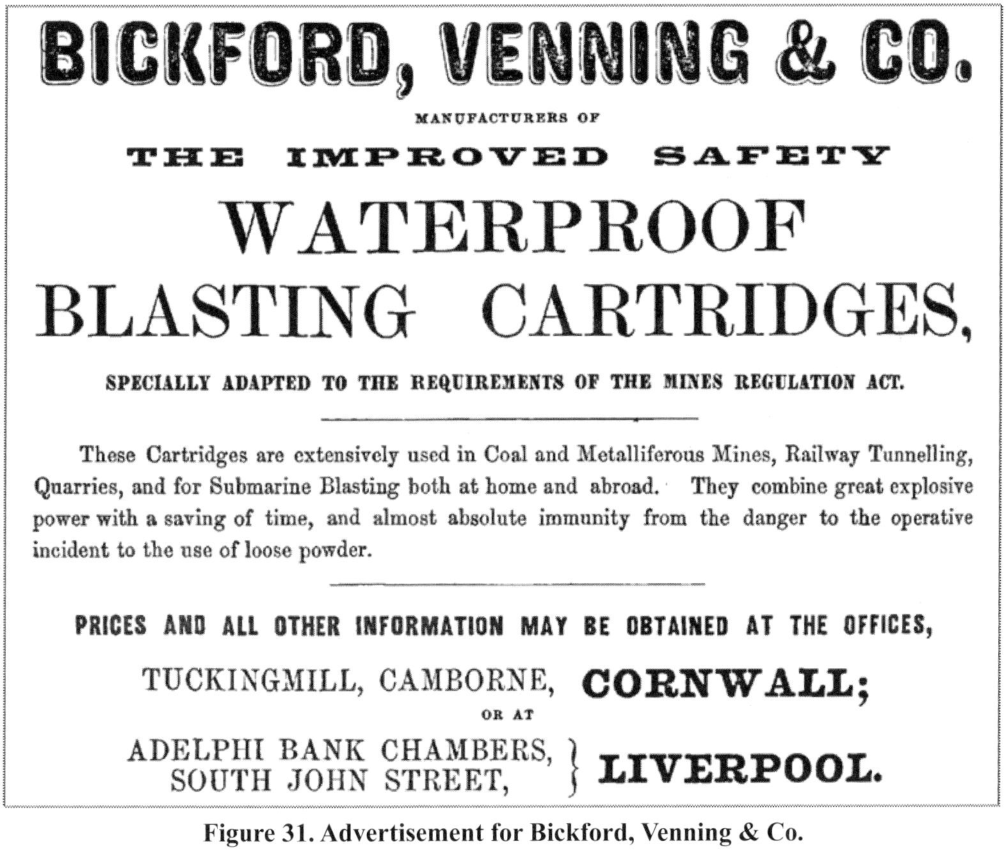

Figure 31. Advertisement for Bickford, Venning & Co.

married the sister of Sir R. R. Vyvyan in 1832, is believed to have lived in Antron Lodge in Sithney (which he enlarged) while the new mansion was being built.

In June 1875 the Explosives Act was passed, and came into operation on January 1st 1876. This Act amended the Law with respect to manufacturing, keeping, selling, carrying and importing gunpowder, nitro-glycerine and other explosive substances. Extremely strict rules were introduced for factories handling gunpowder and storing it in magazines. This was probably the reason Bickford, Smith and Co. started to build at least two gunpowder magazines at nearby Dolcoath at the start of 1876. The *Royal Cornwall Gazette* described them as 'new and commodious'.[5]

A disastrous fire and explosion occurred in 1875 at the Company's fuse factory in Meissen, in Germany. At least eleven people were killed. Freeman Eales took over the factory and became joint managing director with his father Joseph. A salesman at Meissen, a Mr Lyth, was sent to purchase a site in Wiener Neustadt, in Austria. The Company was registered in the Austrian Company register on March 4th 1879. Samuel Hocking, engineer to Bickford, Smith and Co. was responsible for its construction.[6]

In February 1877 Samuel Hocking died at his residence in Lower Rosewarne, Camborne, aged 70. A short notice in *Bibliotheca Cornubiensis* stated that Hocking

PATENT SAFETY FUSE

For conveying Fire to the Charge in Blasting Rocks, &c.

BICKFORD, SMITH & CO.,

Manufacturers & Original Patentees.

BICKFORD'S PATENT SAFETY FUSES

OBTAINED THE PRIZE MEDALS AT THE

"Royal" Exhibition of 1851;
AT THE
"International" Exhibition of 1862, in London;
AT THE
"Exposition Imperiale" in Paris, in 1855.

AT THE
"International" Exhibition in Dublin, 1865;
AT THE
"Exposition Universelle" in Paris, 1867;
AND AT THE
"Great Industrial Exhibition" in Altona, in 1869.

Every Coil of Fuse manufactured by BICKFORD, SMITH and Co. has **TWO SEPARATE THREADS** passing through the column of Gunpowder, and they claim such **TWO SEPARATE THREADS** as exclusively their **TRADE MARK**.

Every package of Fuse sent out by BICKFORD, SMITH and Co. has a *copyright* label attached, bearing their name and address.

FACTORIES:—

TUCKINGMILL, CORNWALL.

AND

St. Helens' Junction, LANCASHIRE.

Figure 32. Post Office Directory advertisement, 1873.

Figure 33. Trevarno in the early years, c.1880. PB-S archives

had constructed machinery for Bickford, Smith and Co. in Lancashire, France, Prussia, Spain and America. Another engineer for the Company, William Bennett, had left to start a rival company in nearby Roskear sometime in 1870.

In February 1879, George J. Smith and William Bickford Smith took out a patent for 'An instrument for the simultaneous ignition of a number of fuses'. This proved to be a very successful product for the firm, and a drawing of it was added to their price list.[7]

The Patent Safety Igniter or 'volley firer' consisted of metal tube which contained a compressed pellet of gunpowder. By bringing a number of instantaneous (or very rapidly burning fuses) together in the metal tube (held together by crimping), they could be lit by one slow burning fuse. Each instantaneous fuse led to an individual charge, and the charges could be fired simultaneously by this method. The volley-firer was particularly useful in quarrying, and also in Cornwall in shaft-sinking.

A very full account of the testing was given in *The Engineer* in December 1880.[8] A series of seven patent igniters were fired on Truro Green before the Mining Institute of Cornwall. Each igniter contained from four to twelve instantaneous fuses. Some of these were of the quality prepared for dry soil; one contained the large and well-coated fuse for underwater blasting; but the majority were made up with a neatly covered tape instantaneous fuse, such as would be commonly used in ordinary mining and moderately

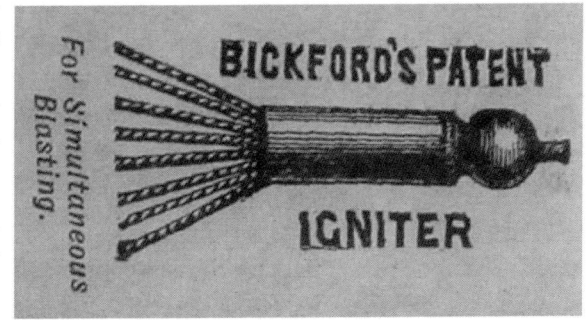

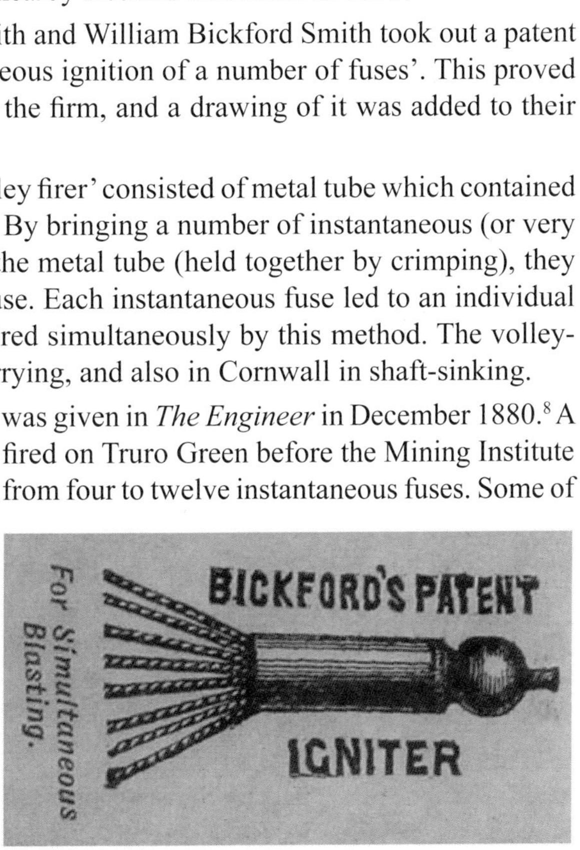

Figure 34. The patent *Igniter*.

damp places.

In each case the speed of the explosion was equal, the only difference being that the more heavily coated fuses produced a louder report that those more lightly made and designed for dryer ground. In each case it was obvious that had the instantaneous fuse in question, whether four or twelve, had each been attached to a charge of explosive, the result would have been a simultaneous blast of the whole number of charges, an effect which is found important in many driving operations, especially where the boring machines are employed. It was noted that the cost of these fuses was considerably greater than the ordinary safety fuse.

The experiments concluded with the burning of a piece of instantaneous fuse more than 100 feet (30 meters) long, 'the intensely rapid flash of which caused no little sensation amongst the crowd', and tended to confirm the scientific tests that its

Figure 35. Price list c.1892

speed exceeds 100 feet per second. George J. Smith conducted the experiments, and the Institute awarded Bickford, Smith and Co. its silver medal.

Price List

The company's price lists began to reflect the various medals they had won in exhibitions throughout Europe. It important to remember that at this time the promotion of their fuse was entirely down to the Company itself – there was no assistance from any Government department, as is the case today. The list of exhibitions entered and medals won was most impressive.

London	1851	Marseilles	1880
Paris	1855	Liverpool (Gold)	1886
London	1862	Newcastle	1887
Dublin	1865	Adelaide	1887
Paris	1867	Brussels (Gold)	1888
Altona (Silver)	1869	Vienna	1888
Cordova (Silver)	1871	Melbourne (Gold)	1888
Vienna	1873	Glasgow	1888
London	1874	Barcelona	1888
Santiago	1875	York (Gold)	1889
Philadelphia	1876	Paris (Gold)	1889
Capetown (Gold)	1877	London Mining (Gold)	1890
Sydney	1879	Dunedin	1890
Truro (Silver)	1880	Edinburgh (Gold)	1890
Melbourne (Silver)	1880	Plymouth (Gold)	1890
Boston (Jewel)	1883	Leeds (Silver)	1890
London (Gold)	1884	Sheffield (Gold)	1891
London (Gold)	1885	Tasmania	1891
Paris (Gold)	1885	Birmingham (Silver)	1892

Bickford, Smith and Co.'s price list was comprehensive, and was accompanied by a short description of the various coloured fuses (see Figure 116).

No.	Type	Per coil of 24 feet	No	Type	Per coil of 24 feet
1	PATENT SMALL FUSE, for immediate use in dry ground	3d or 3½d	17	BICKFORD'S METALLIC FUSE, for use in very wet ground	1/-

2	PATENT FUSE, for use in dry ground	4d		18	PATENT DOUBLE COUNTERED METALLIC FUSE (White), for use under water	1/6
3	PATENT WHITE FUSE, for use in dry and close places	4d		19	PATENT DOUBLE COUNTERED METALLIC FUSE (Black), for use in deep water	1/6
4	PATENT RED FUSE, for use in very damp and close places	4½d		20	PATENT TREBLE COUNTERED METALLIC FUSE, for use in 40 feet of water, when immersion for a considerable is necessary	2/-
5	PATENT DOUBLE WOVE FUSE, for use in damp ground	4½d		21	PATENT GUTTA PERCHA COUNTERED METALLIC FUSE, for use in any depth of water, and will bear immersion any length of time	2/6
6	PATENT THREAD SUMP FUSE, for use in wet ground	4½d		22	PATENT TREBLE WOVE FUSE, a very superior Fuse, adapted to almost every mining requirement	6d
7	PATENT SMALL TAPE SUMP FUSE, for immediate home use in wet ground	5d		15	PATENT DOUBLE GUTTA PERCHA FUSE, for use in 300 feet of water	1/4 and 2/-
8	PATENT TAPE SUMP FUSE, for use in wet ground	6½d		16	PATENT IMPERMEABLE SUBAQUEOUS FUSE, for use in any depth of water for any length of time, will bear any pressure and very great strain.	from 8/-
9	PATENT DOUBLE TAPE SUMP FUSE, for use in very wet ground	8½d		23	PATENT WHITE TAPE FUSE, for use in wet and close places, and for exportation into warm climates	6½d
10	PATENT DOUBLE COUNTERED FUSE, for use in very wet ground – will bear rough treatment	7½d		24	PATENT WHITE DOUBLE TAPE FUSE, for use in very wet and close places, and for exportation into hot climates	8½d

11	PATENT THREAD COUNTERED TAPE FUSE, adapted for wet ground – strong and tough	8d	25	NEW PATENT COLLIERY FUSE, to convey fire to the gelatinous or water-cartridges, or other charge, without emitting any fire laterally during combustion.	8½d
12	PATENT SMALL GUTTA PERCHA FUSE, for use in very wet ground or water	10d	26	PATENT WHITE TREBLE FUSE, for use in wet ground and any kind of soil; particularly recommended for export to foreign and tropical markets.	7d
13	PATENT GUTTA PERCHA FUSE, for use in 40 feet of water	1/-	27	PATENT WHITE COUNTERED GUTTA PERCHA FUSE, for Home use Ditto for export	8d 8½d
14	PATENT TAPE GUTTA PERCHA FUSE, for use in 60 feet of water.	1/3 and 1/6			

Price List 1892

Under the heading 'A Cornish Trade Mark Case', the *Royal Cornwall Gazette* reported, in July 1882, that Bickford, Smith and Co. had applied to the Courts to force the Swansea Fuse Company to alter their trade mark.[9] This company's trade mark consisted of a goat with "The red thread" through the centre of the fuse. Bickford, Smith and Co. alleged that since 1865 they had used two threads, one white and the other red. However, they asserted that the red thread was the distinguishing mark, and their fuses were known by it. They wanted the Swansea Fuse Company to remove the reference to 'the red thread'. The judge, Mr. Justice Chitty, refused their application.

The Tuckingmill premises clearly needed enlarging, and in March 1882 an advertisement was placed in the *Royal Cornwall Gazette*. Tenders were invited from local builders 'for the erection and completion of offices etc.' Tenders were to be delivered to the architect, James Hicks, by March 18th 1882.[10] Hicks (1846-1896) was a well-known architect at the time, and designed several notable houses, schools, chapels and churches in the Camborne and Redruth area. He was appointed as Chief Agent for local landowner Lord Clinton, and developed Clinton Road, Redruth. Here he designed not only the Passmore Edwards Library but also other notable buildings, one of which was his own house.

The contract was awarded to Alfred Jenkin, from Leedstown. During the same period Jenkin was also awarded the contract to build Porthleven Chapel. He may have overstretched himself, and became bankrupt.

In November 1883 an interesting case involving the building of the new offices

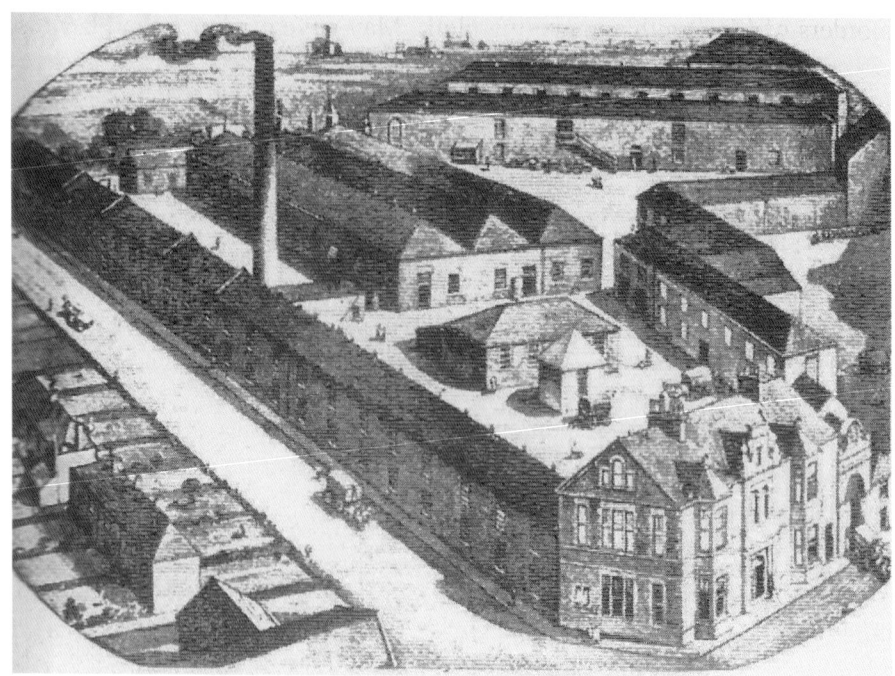

Figure 36. Letterhead from the late 1880s

came before Judge Montague Perk, Q.C., sitting in Truro County Court.[11]

Trustees in bankruptcy of Alfred Jenkin of Leedstown, applied to the Court for an order to get Bickford, Smith and Co. to pay £200 to the estate.[12] The fuse company had given Alfred Jenkin the contract for the erection of their new offices, at an agreed price of £1,301. After completing quite a lot of the work, Jenkin received £920, but got into financial difficulties. The Gweek Trading Company agreed to take over the contract and guarantee completion. At this time, the work remaining was valued at around £150, and Jenkin expected the full amount of the contract money would be paid, plus another £75 for extra work. Smaller contractors, named as Messrs. J. and C. Tyack, W. Rabling and Co. and Mr. Chynoweth of Camborne were also owed money. Bickford, Smith and Co. was ordered to pay £193 into court. They had already paid the Gweek company £180. The case came back into court in February 1884, where Bickford, Smith and Co. applied to have £60 returned to them, under a penalty clause for late completion of the work. It seemed that the Gweek company, which took over the original contract, agreed to the penalty clause contained in it. They agreed to have the work completed by March 1st, 1883, but did not do so.[13]

William May, a builder from Pool, testified that when he had started on February 1st, 1883, the Gweek Company was involved in completing the work. May stated that the extra work interfered with their contract work, adding about a fortnight to a month to the time allowed, and that the extra work cost about £30. Bickford, Smith and Co. supplied the mantelpieces, some of which had to be taken down and altered, on the orders of Mr. Smith's managing clerk. May gave evidence that, because the offices were occupied, the plasterers were greatly delayed. The clerks took possession upstairs before the oak staircase was finished, hindering the works. He added that the banisters were 'turned' in Truro. May said that the work was completed by April 25th 1883, and he informed Mr. Jeffery, of the Gweek Trading Company.

At the resumption of the case, in March 1884, John Haslett, a plasterer from Camborne, testified that he had finished his work by the end of March or the start of April. Mr Chynoweth, a Camborne painter, gave evidence that he had finished by the end of April, although delayed by all the extras demanded by Bickford, Smith and Co. The fuse company alleged that the Gweek Company was not off the premises until June 24th, while the Gweek company stated that it had finished by the end of April, and that the delays were caused by the extras ordered by the fuse company, and also by Bickford, Smith and Co. taking partial possession of the premises. Mr Jeffery, of the Gweek Trading Company, told Judge M. Bere that architect Hicks had informed him that the work would cost £160, but instead it had actually cost £383. The judge replied that if the Gweek Company took the contract without any agreements in writing, then that must not interfere with the original contract. The only question was whether Bickford, Smith and Co. could claim penalties if they themselves broke the agreement by the delays. George J. Smith gave evidence that the Gweek Company knew that the offices were partially occupied when they started work. James Hicks, the architect, said that both he and the original builder, Alfred Jenkin, thought that the work would be completed by March 1st, 1883. He inspected the work every three to four weeks, and thought he was last there on the 23rd of May, 1883, when he found fault with some unimportant items.

On May 15th 1884, Judge Bere gave a decision in favour of the trustees in bankruptcy of the builder Alfred Jenkin.[14]

A letterhead survives from an exhibition in the 1880s. It shows a large fuse factory, with the front façade

Figure 37. The interior of the Porthleven Institute 2013

resembling the current buildings. This accurately shows the buildings drawn on the 1880 Ordnance Survey map, but the fine office building on the corner of Dolcoath Road (or Row) and Pendarves Street shows the new offices designed by James Hicks. The entrance with arch over is clearly shown in the drawing.

In October 1884, Bickford, Smith and Co. applied to the Home Office for an amending licence. This would have been mandatory under the 1875 Explosives Act every time the factory was enlarged. This was the fifth amending licence that they had applied for. The first application was in September 1876, followed by others in January 1878, November 1878, and October 1880.[15] It can be seen from the drawing just how cramped the site had become. The acquisition of the adjoining Tuckingmill Foundry was some years off.

The County Record Office in Truro has a plan to accompany the amending licence application of 1884. It is annotated, and the buildings are numbered and listed.

In December 1884, the Porthleven Institute was officially opened by William Bickford Smith of Trevarno. It had been his personal gift to the town, and stood on the site of the old Fisherman's Arms. The same architect, James Hicks of Redruth, who had designed the fuse factory's offices, also designed the Institute. The clock, which had four dials, had been started in November by Mrs. Bickford Smith, and had been made by Benson at a cost of £200.[16]

In 1885, William Bickford Smith became the Liberal MP representing the Truro-Helston Division, and shortly afterwards changed his name by deed poll, joining his middle name and his surname by a hyphen, and afterwards was referred to as William Bickford-Smith.[17] In July 1886, he was again elected, this time as a Liberal-Unionist. This came four months after the death of his mother Elizabeth Smith.

The attention of the fuse company turned to the development of a method of lighting their fuses in gaseous coal mines. The Coal Mines Act of 1887 came into operation at the start of January 1888, and forbade naked flames. Rule 10 stated: - No person is allowed to have in his possession any Lucifer match or apparatus of any kind for striking a light, except within a completely closed chamber attached to the fuse of the shot.

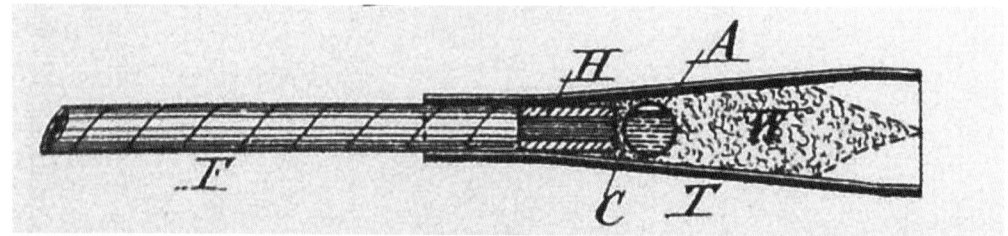

Figure 38. Illustration from the patent application 1887. A is Acid, C is the Chlorate mixture, F is the Fuse, H is the fuse plug, T is the Tube, W is a soft material such as cotton wool filling the tube.

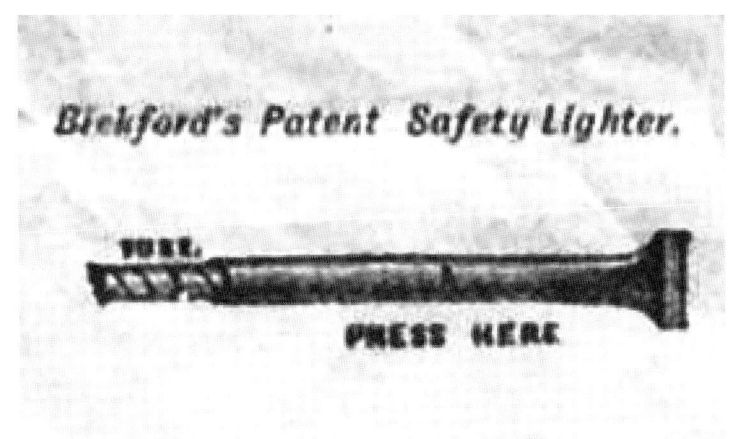

Figure 39. Illustration from Bickford, Smith and Co's price list.

In January 1887 William Bickford-Smith and George Smith applied for a patent for a 'Means of Igniting Fuses without exposing flames or sparks'. It was accepted on November 10th 1887. A pill consisting of a mixture of chlorate and potash and sugar was brought into contact with a globule or tablet containing sulphuric acid in an enclosed metal tube, made of a soft metal such as tin. When the acid came into contact with the mixture, combustion took place within the tube, and the fuse was ignited.[18]

When the end of the fuse was pushed into the tube, the tube was squeezed closed over the fuse, and again pinched in the middle to bring the chemicals into contact one with another.

This was a successful product for Bickford, Smith and Co. and appeared on their price list.

Key to Plan to Amending licence No. 238. 13th October 1884

1	General Office	17	General Store
2	Residence	18	Warehouse
3	Office	19	Mason's Room
4	Powder Dry	20	Soldering Room
5	Warehouse	21	Tar boiling Shed
6. 6	Yarn Dries	22	Carpenters' Shop
7. 7	Boiler House	23	Warehouse
8	Engine House	24	Warehouse
8	Upper Room over Engine House	25	Pitch and Resin Shed
8A	Dry for Impregnated Yarns	26	Waste Room
9,10,11,12	Upper Spinning Rooms	27	Upper Dressing Rooms
9	Lower Fitting Shop	27	Upper Dressing Rooms
10,11	Lower Spooling Rooms	28	Kitchen
12	Lower Instantaneous Fuse and	29. 29	Lower Lavatories

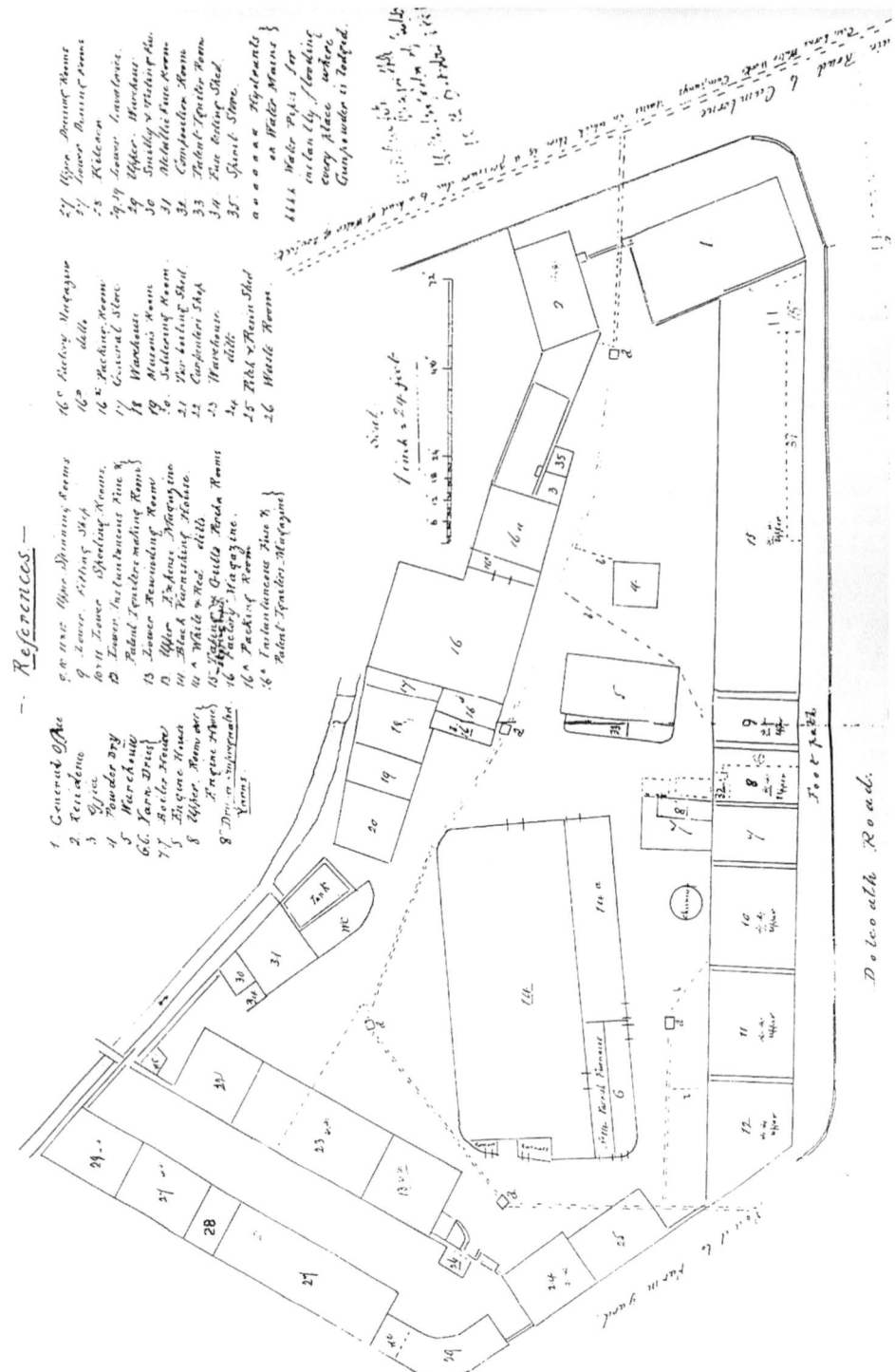

Figure 40. Plan to accompany application for an amending licence No. 238, 13th October 1884

	Patent Igniter making Room		
13	Lower rewinding Room	29	Upper Warehouse
13	Upper Expense Magazine	30	Smithy and Testing Fuse
14	Black Varnishing House	31	Metallic Fuse Room
14A	Red and white Varnishing House	32	Composition Room
15	Taping and Gutta Percha Room and storing tapes	33	Patent Igniter Room
16	Factory Magazine	34	Fuse Testing Shed
16A	Packing Room	35	Spirit Store
16B	Instantaneous Fuse and Patent Igniter Magazine		
16C	Factory Magazine	aaaa	Hydrants or water mains
16D	Factory Magazine	bbbb	Water pipes for instantly flooding every place where Gunpowder is lodged
16E	Packaging Room		

References, Chapter 7

1. Information from Bryan Earl.
2. Advertisement in PO Directory 1873
3. *The West Briton*, July 17, 1873
4. *Royal Cornwall Gazette*, December 5, 1874
5. *Royal Cornwall Gazette*, February 26, 1876
6. See Reference 1, Chapter 3. G. L. Wilson
7. *The West Briton*, October 3, 1879
8. *The Engineer*, December 17, 1880
9. *Royal Cornwall Gazette*, July 21, 1882
10. *Royal Cornwall Gazette*, March 10, 1882
11. *The Cornishman*, November 22, 1883
12. Trustees in bankruptcy are appointed by the courts to ensure that everything the bankrupt is ordered to do is, in fact, carried out. They administer bankrupt estates.
13. *The Cornishman*, February 21, 1884
14. *The Cornishman*, May 22, 1884
15. Document in the National Archives. X 66628
16. *The Cornishman*, December 18 and December 25, 1884
17. He will be referred to in this book as William Bickford-Smith from the date of the deed poll.
18. Document in Peter Bickford-Smith Archives.

Figure 41. Mrs. Elizabeth Smith. All PB-S archives

Figure 42. Sir George Smith

Figure 43. William Bickford-Smith

CHAPTER 8

Bickford, Smith and Co. Ltd.
1889-1899

On January 1st 1889, Bickford, Smith and Co. became a limited liability company. The Company presented for registration on 29th December 1888 through their solicitors Ingle, Cooper and Holmes, 20 Threadneedle Street, London. The Certificate of Incorporation was received on January 1st 1889.[1] The Nominal Capital was £130,140, divided into shares costing £20 each.[2] William Bickford-Smith (who seemed to be in poor health) gave written authorisation to the Bickford Smith Company Secretary, Vivian Pearce, to act as his agent in signing the Memorandum and Articles of Association. The Memorandum stated that the Company was in the joint-ownership ('co-partnership') of William Bickford-Smith, George John Smith and Simon Davey at Tuckingmill, St. Helen's Junction and 'elsewhere'.

The seven men who set up the Limited Company, known as 'subscribers' were either descendants of William Bickford and his partner Thomas Davey, or had married into Simon Davey's family in France. At this time they took one share each.

1.	William Bickford-Smith	Trevarno, Helston	Manufacturer
2.	George J. Smith	Treliske, Truro	Manufacturer
3.	S. Davey	Rouen, France	Manufacturer
4.	H. Arthur Smith	Lincoln's Inn, London	Barrister-in law
5.	James Watson	Rouen, France	Manufacturer
6.	A. Harlé	Rouen, France	Manufacturer
7.	Charles E. Tyack	Trevu, Camborne	Accountant

The original directors were William Bickford-Smith, George J. Smith, Simon Davey, James Watson and Alfred Nobel. James Watson replaced his brother-in-law Simon Davey when the latter died on April 29th. The fuse company clearly needed funds to expand, and these were procured by becoming a Limited Company.

The directors turned their attention to expanding into Australia. In 1888 Bickford's purchased the fuse factory of Charles Perry and his nephew John Hunter in Bendigo,

Victoria.[3] They had established the factory around 1875, while manufacturing stamper gratings used in quartz crushing machines on the Bendigo gold field. Additions were built, and it seems that Bickford, Smith and Co. had the Cornish Shield, with its fifteen bezants, carved into the gable end of the building, with the words ONE AND ALL carved on a scroll underneath. When mining declined on the gold field, around 1912, the factory closed.[4]

In 1907 Nobel's and Bickford, Smith and Co. bought a half share each in a fuse factory Charles Perry had started at Footscray, near Melbourne, paying £14,000. Fuse prices were then raised, and profits doubled to over £4,000 per year. Bickford's replaced the Bendigo fuse works in 1911 with a new factory at Spotswood, south west of Melbourne, which closed after World War I.[5]

1889 saw the death in Rouen of Simon Davey, the son of William Bickford's original collaborator in making the first fuse spinning machine. His personal estate was valued at £96,000. He bequeathed 1,000 £20 shares to his son Eugene Henri, and 200 shares to another son Francis. The rest of his property he left to his wife, and also to his son Francis and daughter Alice Harlé.[6]

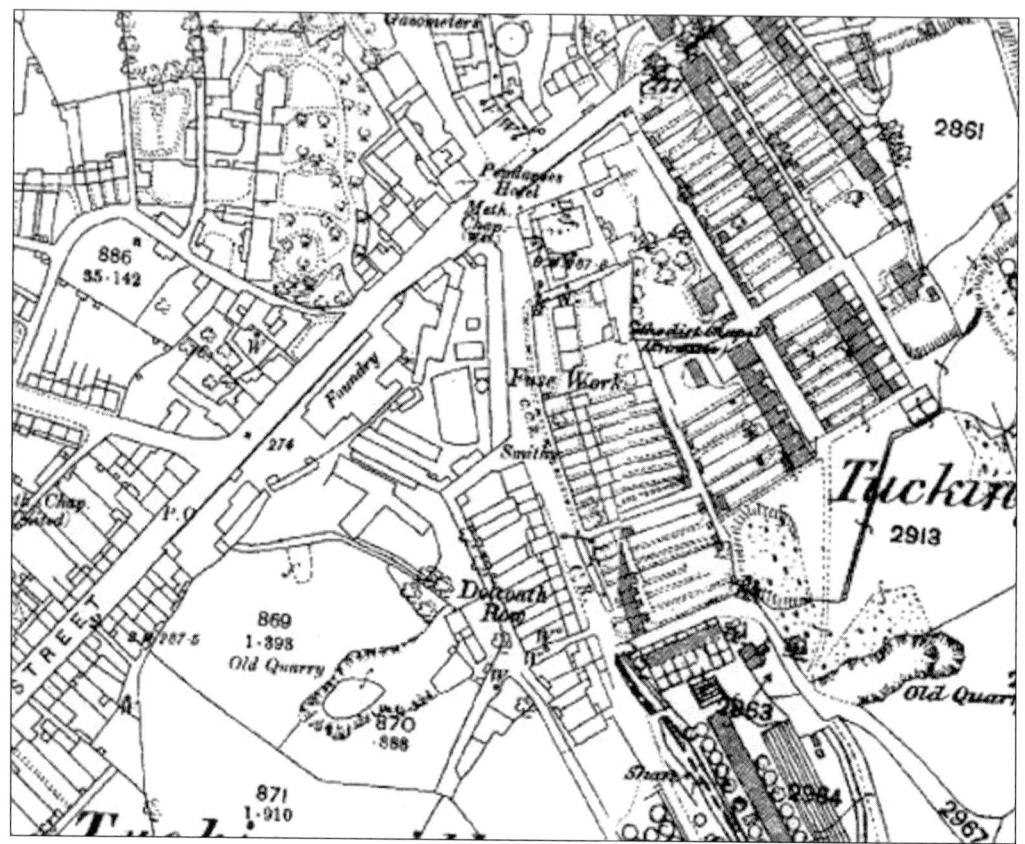

Figure 44. The factory in 1888. Expansion was possible to the south and west.

In Cornwall, the fuse factory in Tuckingmill needed to expand. As it was hemmed in on two sides by Dolcoath Row (now Chapel Road) and by Pendarves Street and the Tuckingmill Foundry, it could only expand southwards, and also to the west. The number of people employed at the fuse factory around this time is not known, although some idea can be estimated by the reporting of the Bickford, Smith and Co. annual outing on Wednesday, August 17th, 1892. Apparently seven horse-drawn brakes and 21 horses took the workers to Carbis Bay and back.[7]

Production of safety fuse in 1895 was as follows: -

Tuckingmill: 1.23 million coils,
St. Helen's: 440,000 coils,
Bendigo: 350,000 coils.

In 1875, the Tuckingmill factory alone produced this total number of coils, but competition affected their sales.[8]

The factory was busy manufacturing patent safety and instantaneous fuse, volley firers and safety lighters on what was a very cramped site. In July 1896 there was a small report in *The Cornishman* to say that the enlargement of the factory was taking in all the old road leading up to Charles Mayne's farmyard, and also part of Mr. Murdock's offices. The mason was T. Turner, and the newspaper reported that the work was giving employment to a great many, and 'was quite a Godsend in these depressed times'.[9] The factory expanded south, into an old quarry. The new buildings included a southern range, which included lavatories, work-people's dining rooms, workers' dressing rooms, a kitchen, a second dining room and more lavatories. Nothing remains of this range – a small housing estate has been built where it once stood. In August, Bickford's asked if Camborne Urban District Council could asphalt the path from the east end of their premises, to which the Council agreed.[10]

In 1896, Bickford, Smith and Co. started a steel works at the old jute factory in nearby Pengegon. Six men were employed to make furnaces and steel for the Tuckingmill Foundry. Apparently old rivets and pig iron were converted into steel.[11] Bickford's interest in the Pengegon works may have come about because of their interest in the Tuckingmill Foundry. In February 1893, the Tuckingmill Foundry and Rock Drill Company was registered as a Limited Company, with capital of £50,000 in £5 shares. The owners at that time were George J. Smith, William Bickford-Smith, H. Phillips Vivian and J. R. Daniell. The Registered Offices were 1 Chapel Street, Camborne. Nothing more about the steel works appeared in the local newspapers.[12]

In December 1896, the Home Secretary, under Section 6 of the Coal Mine Regulation Act, passed an order banning any but a permitted explosive in any part of a coal mine, to come into effect on January 1st 1898. He stipulated that the apparatus or method used, whether electrical or otherwise, in igniting the charge, should be 'as far as is reasonably practicable, be incapable of igniting gas or coal dust. It seems that the expansion of the Tuckingmill factory may have been necessary to meet the demand for the new safety lighters.[13]

There must have been great excitement in the fuse works in the early summer of 1897. The *West Briton* newspaper in June ran a headline which proclaimed: 'A NEW CORNISH KNIGHT', accompanied by a photo of Sir George Smith in his military uniform, probably that of the 1st Volunteer Battalion of the Duke of Cornwall's Light Infantry, of which he was a Colonel.[14]

The newspaper noted that in Queen Victoria's 60th Jubilee Honours List, it was the only honour to go to Cornwall.

> The fact that it has fallen to the lot of Mr. Smith will be a source of gratification to his fellow-Cornishman generally, and to the Non-conformists of the county in particular.

Two weeks later, on July 8th, Bickford, Smith and Co. Ltd. had their annual outing. 150 employees and their wives were conveyed in seven brakes and 26 horses, all bedecked in Jubilee colours, to the beautiful beach at Praa Sands. When Sir George's carriage arrived, he received 'three cheers' from all the employees.[15]

The expansion of the Tuckingmill factory had obviously been extensive, doubling the factory in size from 1884. By this time they were manufacturing patent safety and instantaneous fuse, volley firers and safety lighters. An amending licence (no.767) was applied for, to the Home Office, at the start of October 1897. The plans of the fuse factory to accompany this licence are missing, but a description of all 67 rooms can be found in a document in the National Archives in Kew, London.[16] (See Appendix 1) However, there is what appears to be a traced plan, in pencil, on fragile paper, in the Cornwall Record Office. It is slightly later than the 1897 amending licence, and is dated in pencil '1908', but the marked and numbered rooms seem to match the description found in Kew, apart from a few buildings added after 1897. (See plan in the Appendix 1). When the amending licence (no. 850) was finally granted on September 1st, 1899, an unknown person had written on the front: - 'Express hope that particulars and definition of "Safety Instantaneous Fuse" will be deemed a confidential memorandum of the

Figure 45. Sir George Smith

Department, as otherwise it would be injurious to their interests.' There follows a brief description on the reverse, as written by Bickford, Smith and Co. – '(it) does not contain its own means of ignition, and of which the coating is of such strength and construction that the burning of such fuze will not communicate laterally with other fuzes'.[17]

It seems that the inhabitants of Dolcoath Row, and the Lord of the manor (Mr. Pendarves), allowed the factory all the scope they needed to expand the works. In return, Bickford, Smith and Co. Ltd removed the dangerous old windlass which the inhabitants were using to draw water, and replaced it with a shoot-pump.[18]

Bickford, Smith and Co. Ltd was facing great competition at this time. It is thought that about 10% of the British market in 1895 was going to imported German fuses. The British fuse factories at that time were:-

Roskear, Camborne	W. Bennett
Penhellick, near Tuckingmill	W. Brunton
Brymbo, Wrexham, N. Wales	W. Brunton
Redruth	E. Tangye
Redruth	British and Foreign
Scorrier, Redruth	Unity Fuse
Swansea, Wales	Swansea Fuse Company

Swansea Fuse Co., Swansea, Wales
Nobel's 'Thistle Brand' was not manufactured by them, but was bought from the Swansea Fuse Company. They bought the Swansea Company outright in 1896. Nobel's Explosives Company were also buying almost one million coils annually from Bickford's, but sold it under their own label, apparently at Bickford's insistence. In 1899 Nobel's bought Tangye's of Redruth, closed it, and then moved the machinery to their Swansea Company. Bickford's also disposed of some of the local competition by buying the nearby fuse works at Penhellick, Pool, and closing it at the end of 1898.

In 1897 George Percy Bickford-Smith made an application to the Home Office to erect a safety fuse factory at Trevarno. It was intended to manufacture electric detonators there.[19] It was reported that lack of sufficient space at Tuckingmill was now a drawback for the expanding factory there, and this caused a site at Trevarno to be considered. It seems work may have started at the Trevarno factory, but the death of, firstly, William Bickford-Smith in 1899, followed by the death of his son Percy in 1901, seems to have completely halted the project.

William Bickford-Smith died at Trevarno on 26th February 1899. The *Royal Cornwall Gazette* reported that he 'took to his grave the marks of a hero', with his

Figure 46. William Bickford-Smith. PB-S Archives

face badly scarred by fire.[20] He had apparently rushed into the burning fuse factory at Tuckingmill when younger, and rescued several female employees, during which act of bravery he suffered severe burns. He left a personal estate valued at £47,000 and the gross value of the whole estate was £80,579.[21] Two years later, on 30th May 1901, his son George Percy died in Heilbron, fighting in the Boer War in South Africa. He was a Captain in the 1st Volunteer Battalion Duke of Cornwall's Light Infantry. There is a plaque in his memory in Chynhale Methodist Chapel, near Trevarno, Helston.

References, Chapter 8

1. Company Registration No. 27944
2. Cornwall Record Office. BY/353, Nat Archives BT/31/31130/27944
3. G. L. Wilson. See Chapter 3, reference 1.
4. Document in Peter Bickford-Smith Archives
5. G. L. Wilson, *op. cit.*
6. *The Morning Post*, August 5, 1889

7. *The Cornishman*, August 18, 1892
8. G. L. Wilson, *op. cit.*
9. *The Cornishman*, July 10 1896
10. *The Cornishman*, September 3, 1896
11. *The Cornishman*, August 27, 1896
12. *IRON*, February 24, 1893
13. *The Yorkshire Post*, December 23, 1896
14. *The West Briton*, June 24, 1897
15. *The West Briton*, July 8, 1897
16. The National Archives, Kew, London. X66623
17. The National Archives, Kew, London. X71.188B
18. *The Cornishman*, November 4, 1897
19. *The Cornishman*, November 4, 1897
20. *Royal Cornwall Gazette*, March 2, 1899
21. *The Cornishman*, May 18, 1899

CHAPTER 9

Bickford, Smith and Co. Ltd. - The Sir George J. Smith Years
1900-1921

When William Bickford-Smith died in 1899, the Managing Director of the fuse factory was his younger brother Sir George John Smith (1845-1921). He was to oversee very far-reaching developments for the company. Sir George's son, Colonel George Edward Stanley Smith (1873-1950) was also working for the company. He served with the Duke of Cornwall's Light Infantry for a greater part of the Boer War. On his return he took over the management of the foreign factories, reorganising and extending them to meet the increased demand.[1]

The works manager, Charles Edward Tyack (1862-1931) was also a Director, and a grandson of Dr George Smith.

Figure 47. Colonel George Edward Stanley Smith DSO

Figure 48. Charles E. Tyack

When William Bickford-Smith died, his twin sons by his second wife Anna Mathilda Bond were only 18 years of age. John Clifford Bickford-Smith and his brother William Noel Bickford-Smith would eventually take over the management of the factory.

'WAR NO FRIEND TO SAFETY FUSE'
In March 1900, *The Cornishman* newspaper ran this rather alarmist headline.[2] The

article stated that, because of the Boer War, 'thousands' of Cornish miners working on the Rand goldfield had been thrown out of work. As a result, work supplying them with fuses had slowed down at Bickford, Smith and Co., although the factory had not reduced their workforce. Bickford's had already reduced some local competition by buying and closing the Brunton's factory (Penhellick, Pool). Apparently the Company approached Nobel's Explosives Company Ltd to form some kind of association to prevent price-cutting, and a Convention of British Fuse Interests was proposed in 1900, but did not progress. It seems that the British and Foreign Fuse Company in Redruth was against the idea.[3]

Sir George Smith, the managing proprietor of the Tuckingmill Foundry and Rock Drill Company Ltd., had to deal with the outcome of a major fire at the Foundry in March 1900.[4] A serious fire had broken out in the pattern shop, where wooden patterns were stored. These were kept in case orders for replica machinery, previously made at the Foundry, were received. The building was L-shaped, each wing being 17 yards in length. One wing was a workshop, while the other was the pattern shop. The fire caused major damage to the Foundry. Because it was so close to the fuse factory, this must have been a serious concern, and may have had a bearing on the decision to close the Foundry some years later. The Foundry was also facing intense competition from Holman Bros. much larger engineering works in Camborne.

Parallel to the development of the Tuckingmill fuse business was the expansion of Alfred Nobel's dynamite empire. He had established the British Dynamite Company in 1871, and the first factory was built on the Ardeer peninsula, Ayrshire, Scotland. In 1886 The Nobel-Dynamite Trust Company was formed, and in 1900 Nobel's Explosives Company Ltd. was registered. Nobel's opened a fuse factory at Linlithgow, in West Lothian, Scotland in 1902. A top executive, Harry McGowan, organised the merger of almost all the explosives and fuse companies after the 1914-1918 War, a move that that included Bickford, Smith and Company Ltd.

Some interesting information about the leases taken by the Bickford, Smith and Co. Ltd factory in 1902 is given in a letter in the Record Office in Truro.

1. Factory. Lease from Mr. Basset. 60 years from 1/1/1902. High Rent £67.
2. Magazine. Lease from Mr. Basset. 60 years from 1/1/1901. High Rent £6.
3. Stores. Lease from Mr. Pendarves. 80 years from 4/1/1902. High Rent £3.

Locally, the tramway system was extended to Tuckingmill in 1902, to the benefit of the large numbers working in the two fuse factories of William Bennett and Bickford, Smith and Co. Ltd.[5] 250 employees of Bickford's enjoyed the annual factory outing to Praa Sands in August 1905.[6] The employees were addressed by Sir George J. Smith every Tuesday at noon, when his talks were mainly lessons from the Bible.[7]

Meanwhile the consolidation of Bickford's fuse business continued. In 1904 Bickford's, together with Nobel's, bought the British and Foreign Fuse factory in Redruth. Nobel's purchased one third, while Bickford's bought two thirds. Two years later they bought Nobel's share. In 1906 Bickford's acquired two thirds of the Unity fuse factory near Scorrier.[8]

A Fatal Fire
The fact that the Bickford's workforce was engaged in a hazardous undertaking was made clear on Friday April 20th 1906, when a fire broke out in a fuse spinning room in which five women were working. They were:-

Jessie Williams, Troon. Uninjured.
Mary Ellen Richards, Tregajorran. Slightly burnt.
Ada Collins, Tregajorran. Slightly burnt.
Lily Sprague, Tregajorran. Badly burnt.
Elizabeth M. Rodda, Camborne. Seriously burnt.

Doctors Gardiner and Blackwood ordered Misses Collins, Sprague and Rodda to be taken to the Women's Hospital, Redruth, while PC Lemin of the Redruth police notified the hospital to make arrangements to receive the badly injured women. J. W. Lean, an accountant at the factory, and local secretary of the St. John Ambulance Brigade accompanied the women, while Sir George Smith arrived at the hospital around 5pm.[9]

Elizabeth May Rodda, aged 48, Centenary Street, died seven days later on the 28th of April. She had worked at the factory for over three decades, and had been present at the fatal explosion thirty-four years earlier, in 1872. Lily Sprague, 21, lived for another four weeks, but died on the 12th of May. Sir George Smith, G. E. Tyack and J. W. Lean attended her funeral in Illogan Church.

The inquest was held on May 15th in Redruth, and was attended by Captain A. P. H. Desborough, H.M. Inspector of Explosives.[10] He noted that the fire was in room 20, which was an upper storey where safety fuse was made. This was one of the upper storey rooms in the long building running down the west side of Dolcoath Row (now Chapel Road).

Figure 49. J. W. Lean

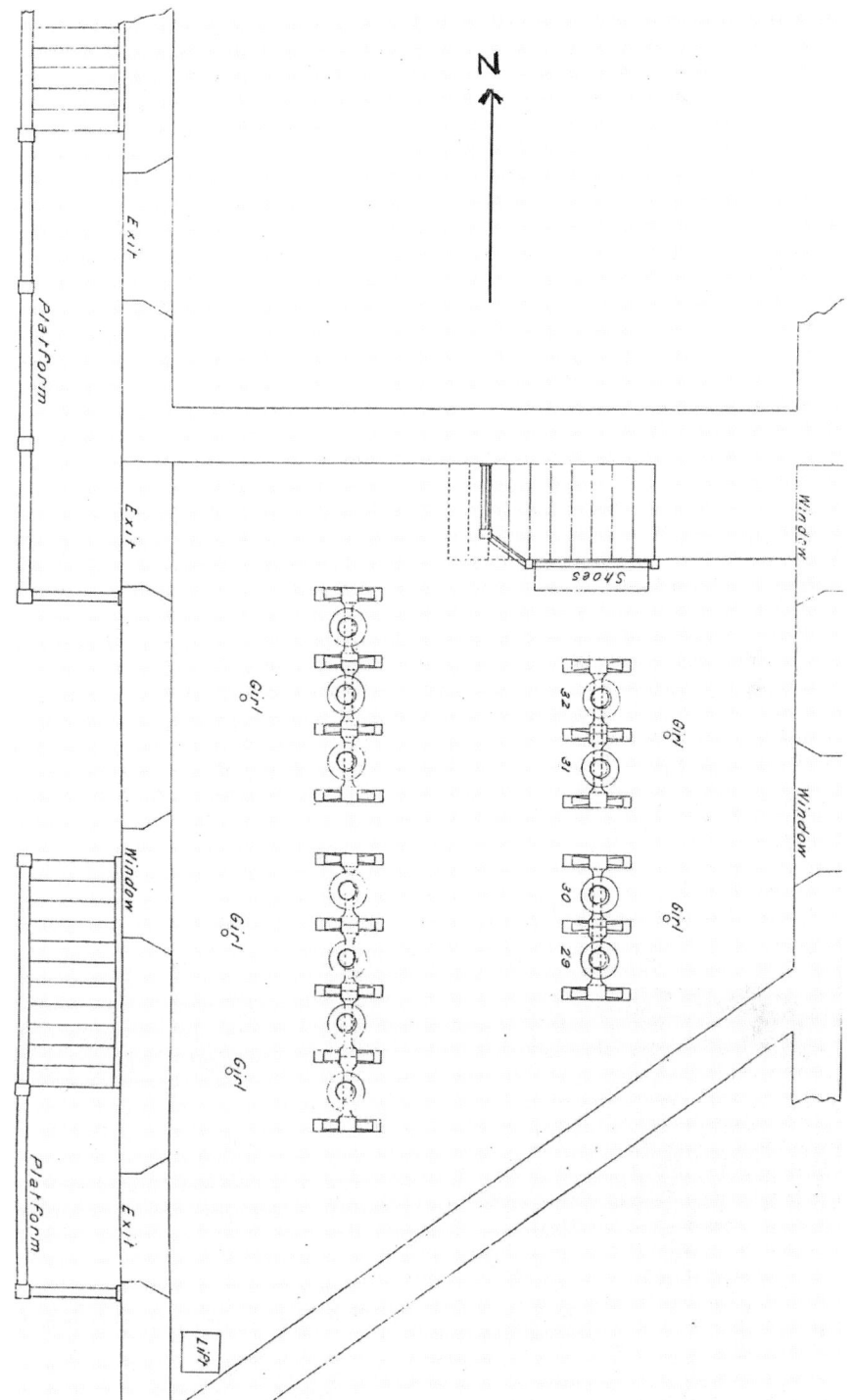

Figure 50. Plan of Room no. 20. The scene of the fatal explosion at Bickford, Smith and Co. Ltd.

Captain Desborough wrote in his report that 'contrary to the usual custom', Room 20, which was licenced for a "danger operation" was not on the ground floor, and that the building of which it formed a part was not of one storey only, as was the usual requirement. He noted that this was really a technical reason. When the factory had applied for their licence, under the terms of the Explosives Act 1876, because they were an existing factory, the terms of the licence were framed to allow as few alterations to the buildings as possible, provided that reasonable security was put in place for the workforce. An upstairs room was sanctioned for 'dangerous work' only if the means of escape was exceptionally good.

As an outcome of the 1872 accident, Bickford, Smith and Co. Ltd had built outside wooden platforms which could be stepped onto through French windows and from there to stairs leading to the ground. Fire hydrants, according to Desborough's report, were working within a minute of the outbreak.

Captain Desborough attached a plan of the spinning room to his report. There were four machines on the east side of the room, in groups of two. On the other side of the room there were seven machines, divided into two groups of two, and one of three. He wrote that it was only after he had searched the room and talked to two of the injured girls was he informed that a steel knife had been found under machine 31, the machine of Lily Sprague. He then discovered that the chief engineer had noticed the knife when entering the room immediately after the fire, and had removed it. He noted that the engineer was possibly influenced by the fact that its presence was a breach of Factory Rules.

> Rule 40. The girls working in the fuze-machine room are to use bronze knives...., the tools are not to be kept on the machines, but in the window or seat behind the girls.

The report continued: - 'If we assume that in all probability the knife did fall and that in so doing it struck the cast-iron framework of the machine, which I noticed was bare of paint in a few places, it might have caused a spark which would be quite sufficient to account for the ignition of the powder dust on the floor.'

His conclusions were:

1. The powder dust on the floor was most probably ignited by the fall of the steel knife
2. The knife had probably been placed on the framework of the machine and was jarred off by the vibration of the machinery.
3. The somewhat abnormal atmospheric conditions had rendered the jute dust and gunpowder unusually sensitive to ignition.

The presence of the steel knife was unquestionably a breach of Special Rules, and moreover, the management of the factory had provided bronze knives for the use of the workpeople, so that at first sight it would appear that a very serious illegality had been committed. Practical experience has, however, conclusively proved that it is impossible to maintain bronze knives in sufficiently good order for the continual use in a spinning room, and that it is necessary to use a harder metal for the blades of knives employed for this purpose. ...I cannot, therefore, say that I think the workpeople were blameworthy except for committing a breach of a rule which, at any rate so far as the actual use of the knife was concerned, was practically impossible to carry out; as regards placing the knife on the framework of the machine I can only say that it was natural for them to place it where it would not be so visible as if placed on the bench. As regards the management, the utmost that can be alleged is that they made an error in judgement in not recognising that a bronze knife was unsuitable and that it would have been wise, therefore, to have allowed the use of a steel instrument under restrictions. That they took this view was probably owing to the provisions contained in Order in Council No. 2, which require that no exposed iron or steel should be allowed in any danger building.

As a tribute to the care with which these works are managed, I may state that there has been only one other fatal accident during the past seventy five years.

Some of Bickford's fuses seemed to have been faulty or substandard in some way, as would be expected in such a large factory. Some idea of how they were disposed of can be gleaned from a newspaper report of June 1906. One of their workmen, Henry Nicholls, was severely burnt about the face and hands when burning waste fuses on the Tuckingmill mine burrows. This was the second such accident in a short space of time.[11]

Permitted Igniter Fuse

In January 1907, Bickford, Smith and Co. Ltd received the welcome news that the Secretary of State had placed Bickford's igniter on the 'permitted list', under the new designation 'Permitted Igniter Fuse'.[12] Ordinary safety fuse was prohibited in 'gassy' mines, namely collieries. The description stipulated 'That the same has been made at the works of Messrs. Bickford, Smith and Co. Ltd at Tuckingmill, Cornwall; or at their works at St. Helens, Lancashire.' The 'permitted igniter fuse' carried the

Figure 51. Bickford, Smith and Co. Ltd Permitted Igniter Fuse.

requirement that the fuse be of special quality, and should be supplied permanently attached to its igniter, with the joint firmly fixed in place by a tape bearing the 'Crown' and the letter 'P'.[13]

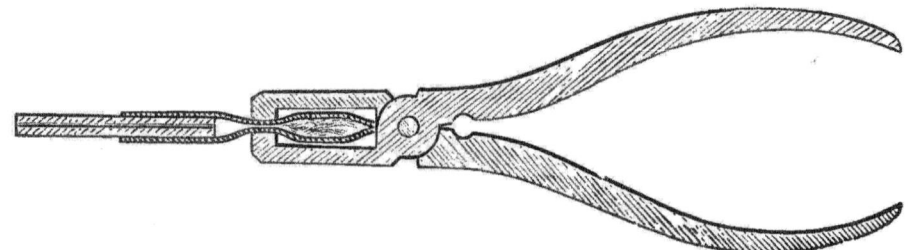

Figure 52. Section through Bickford's Patent 'nippers', with igniter inserted.

The fuse had to be fired by an implement supplied by the manufacturer. Bickford's designed and patented a type of pliers known as a 'nipper'. When the igniter was inserted horizontally for firing, the nipper would close exactly on the right part to squeeze to enable firing. The chamber of the nippers totally enclosed that part of the igniter where ignition occurred, and prevented the escape of any dangerous sparks.

Amalgamation with a Nearby Rival
William Bennett's fuse works at nearby Roskear was the second largest fuse factory in England, after Bickford's. Around 1907, it is believed that Bickford, Smith and Co. Ltd. amalgamated with Bennett's fuse works. The 1909 lists of shareholders for both factories do not show any shareholders in common, but it could be the case that Bickford, Smith and Co. Ltd acquired a controlling interest by buying shares in Bennett's around this time.[14] (See Chapter 12)

Although the two factories were now amalgamated, W. Bennett's always kept their own fuse works name: - Wm. Bennett and Sons Ltd. When a son of the founder (E. J. Bennett) died in 1933, he was a Director of Bickford, Smith and Co. Ltd.

The UK Safety Fuse Trade Association was formed in 1907, and ran for three years. It comprised Nobel's, Bickford's and Brunton's (in Wrexham). Prices were agreed for three years in the UK and elsewhere. Sir George J. Smith was Chairman. Imports from Belgium and German fuse factories were agreed upon, limited to a quota of c 300,000 coils of fuse. Brunton's left the cartel in 1910.[15]

In January 1907 *The West Briton* newspaper printed a small report that Bickford, Smith and Co. Ltd. intended applying to the justices of the East Penwith Division for permission to establish a factory for explosives at Penhellick, Pool.[16] They had bought and closed William Brunton's fuse works there in 1898, but it seems likely that they were considering using some part of that factory. However, three weeks later their solicitor Mr. C. V. Thomas informed the East Penwith Petty Sessions that

Bickford, Smith and Co. Ltd. had applied to the Home Office for an explosive store, which had been granted by the Home Secretary Mr. Gladstone.[17] It seems that the nearby old explosive store of Brunton's remained intact. In addition, there was a gunpowder magazine beside the Penhellick (Pool) works, and the licence may refer to this.

The lack of a footpath outside the factory in Tuckingmill was causing a problem for pedestrians on Pendarves Street. In July 1907, Camborne Council was asked to approve a new workshop at the Tuckingmill Foundry, which was surrounded by the fuse factory. The members thought a committee should be authorised to approach the Company to see if it was possible to obtain the land in front of the proposed building in order to put in a pavement along the side of Pendarves Street. There is no record of the Company's response.[18]

Another small insight into the day to day running of the factory in 1907 was given by an account of an accident involving a truck coming from George Bazeley and Son, Penzance (a steamship company), laden with five tons of pitch. It broke through the roadway into the Red River that was flowing under Dolcoath Row (now Chapel Road). The load was removed and the truck was pulled out of the hole.[19] Bickford's had allocated a shed for pitch, resin and tar on the south side of their site. These were all materials which they used to varnish or waterproof their fuses.

On July 30th 1908, the annual picnic took place at Perranporth. 250 employees left Camborne and Carn Brea stations by special train at 7.45 am. The Company provided them with lunch in the Perranporth Drill Hall, and gave them tea later in the afternoon. Sir George made a short speech to his workers, and the group returned home at 8.45pm.[20]

Charles Tyack, the Manager of Bickford, Smith and Co. Ltd, gave some unique scraps of information to an unknown person on October 4th 1909. The information was written down in a small notebook, which survived the closure of the factory in 1961, and is now in the Cornwall Record Office.

- Wages are 9 pence to one shilling per day.
- 280 girls employed and 12-15 men.
- Extra labour in consequence of the bad arrangement of the works for carrying from one Department to another:-
    5 girls at 1 shilling per day
    2 men at 3 shillings per day
    Total £3 6 shillings per (six day) week.
- In wet weather, goods are liable to damage on transfer from Departments.
- The buildings are more substantial than necessary.
- Geographical position is bad.

- Gunpowder comes from Chilworth in Surrey and Faversham in Kent by rail.
- Yarns from Dundee to Falmouth and Hayle.
- Cotton yarns from Manchester via Liverpool and Hayle.
- Bitumen from London and Liverpool, the bulk from Liverpool and Hayle.
- Goods outwards are chiefly by rail, but sometimes by steamer to Southampton and London, the bulk from London.
- Suggest London as the proper place for their factory.
- All fuse has to be brought to 26A to be examined and then carried to No. 71 to be tarred and then to No. 4 to have the tape put around it. Then from No. 4 to No. 23 to be varnished and coiled, and then back to the packing rooms. After tarring, the fuse is insulated and is then safe.
- The competing factory in England (sic) is Swansea. The only other is at St. Helens and belongs to Bickford, Smith and Co. Ltd. It is now dismantled. The others are in Scotland, one at Linlithgow.
- No. 32A is idle from want of work. It contains 12 machines – 2 are being dismantled.
- Mr. Bickford-Smith says the store which is part of the Foundry was not sold with the Foundry by Sir George G. Smith, who was the owner of the Foundry.
- Sir G. Smith says that some 7 or 8 years ago his son Mr. Stanley Smith surveyed a large number of sites in various parts of England for establishing a new factory, instead of making extensions at Tuckingmill, but the labour question was the factor which determined them not to move.
- The 280 girls employed get 1 hour for dinner and half an hour on Saturdays.
- Sir George Smith says that the wages paid at the St. Helens factory which belongs to him are but a fraction more than at Tuckingmill. Mr. Tyack says that Sir. George refused to give the wages here as he does not think it is a legitimate question.
- Sir George Smith has said he will get out the tonnage in and out, and let Mr. Tyack know.

The Directors of Bickford, Smith and Co. Ltd in 1909 were:-
   Sir George J. Smith. Treliske, Truro. Holding 2550 shares.

   H. Arthur Smith. 2 Elm Court, Temple. E.C. London. Holding 51 shares.

   G. E. S. Smith, Treliske, Truro. Holding 63 shares.

   E. E. Burgess. Melbourne, Australia. Holding 100 shares.

The Nominal Share Capital was £200,000, divided into 10,000 shares of £20 each.[21]

Shareholders in Bickford, Smith and Co. Ltd, 14th May 1909

Name	Address	No. of Shares
Smith, Sir George J.	Treliske, Truro, Manufacturer	2550
Smith, H. Arthur	4 Elm St., London, Barrister at Law	51
Tyack, Charles E.	Trevu, Camborne, Accountant	31
Davey, Madame E. H.	Rouen, France, Widow	980
Davey, Madame R. S.	Rouen, France, Widow	515
Ensign, R. H.	Simsbury, Conn, Manufacturer	100
Ellesworth, L. S.	Simsbury, Conn, Manufacturer	80
Curtiss, C. S.	Simsbury, Conn, Manufacturer	100
Bolitho, J. B.	Trewidden, Buryas Bridge, Tin Smelter	115
Bolitho, J. R.	Trengwainton, Heamoor, Tin Smelter	315
Burgess, E. E.	Melbourne, Australia, Business Manager	100
Pearce, Vivian	Hayle, Accountant	25
Smith, Lady	Treliske	5
Bickford-Smith, Mrs.	Trevarno, Widow	305
Smith, G. E. S.	Treliske, Manufacturer	63
Watson, H. J.	Rouen, France, Manufacturer	10
Harlé, J. A .S.	Rouen, France, Manufacturer	433
Ensign, J. R.	Simsbury, Conn, USA, Manufacturer	75
Bickford-Smith, J. C.	Redbrook, Camborne, Manufacturer	521
Bickford-Smith, W. N.	Trevarno, Helston, Manufacturer	521
Smith, Sir G. J. and G. E. S.	Treliske, Manufacturer	850
Morse, Mrs. S. A. S.	Simsbury, Conn, USA	50
Darling, Mrs. W. E.	Simsbury, Conn, USA	50
Chapman, H. S.	Glenridge, New Jersey, Manufacturer	50
Smith, L. W. B.	Treliske	12
Smith, A. Gordon	Royal Naval Barracks, Chatham, Commander RN	15
Bolitho, R. F.	Ponsandane, Penzance, Banker	50
Bolitho, W. E. T.	Garth House, Penzance, Banker	50

    The Commercial Rates for the towns of Camborne and Redruth, and surrounding area had been assessed by this time.[22] The rates in Camborne were 6s 6d in the pound, while in Redruth they were 6s 2d in the pound. Bickford's were third in rateable value behind Holman's Engineering Works and the Redruth Brewery.

	Old	New (1908)
Holman's Eng. Works, Camborne	255	1440
Redruth Brewery	331	1064
Bickford, Smith and Co. Ltd	203	830
Bennett's Fuse Works, Roskear	120	660
British and Foreign Safety Fuse, Redruth	55	164
Tyack's Hotel, Camborne	93	144
Unity Safety Fuse, Gwennap	40	120
Bickford Smith's magazines, Illogan	12	35

The factory was expanding, and Bickford's was about to embark on an ambitious project to spin their own jute. They had decided on a site on the western side of their site, bordering the Camborne to Redruth road, which at that time ran through Tuckingmill. In 1900, they had produced 2.84 million coils. After the Boer War in South Africa, their production increased to 5.15 million coils in 1905, increasing to 12.09 million in 1910 (including coils produced in Wm. Bennett's factory, Roskear).[23] Nobel's were taking about 1 million coils from Bickford's, and selling it under their own brand. Bennett's had established good overseas markets, including South Africa and Mexico. The Directors clearly felt it was more cost-effective to spin the jute needed in the manufacture of fuses on site, and the Tuckingmill Foundry seems to have been demolished about this time.

Bickford's Sales (inc. Wm. Bennett and Sons Ltd) in 1912[24]

	Million Coils	Value (£ Sterling)
Africa	6.76	85,000
Australia	1.32	29,000
Canada	0.96	11,000
India	0.75	13,000
S. America	0.57	6,000
Mexico	0.54	7,000
Home Market	1.40	25,000
Sundry	0.09	2,000

The question of the Pendarves Street footpath was raised again at a meeting of Camborne Urban District Council in March 1912.[25] Mr. Pendarves, the local big estate owner, had arranged to sell the freehold of the Tuckingmill Foundry to Bickford, Smith and Co. Ltd.[26] The clerk of the Council had written to Sir George Smith to ask if the Company would give the land to the council to build a footpath, which would cost about £120. At the next meeting, in April, Sir George had replied that he needed to consult his fellow directors, some of whom were away.[27] On May 6th Sir George wrote that they were prepared, under certain conditions, to give to the Council the land necessary for the footpath. He gave some details about the Bickford, Smith and Co. Ltd. building work, which suggested they were in the course of erecting what became known as the 'North Lights' building, on Pendarves Street, with its serrated roof This was to be their jute spinning building.[28]

Sir George explained that the Company was erecting a wall from the western end of their property down the hill for 78 yards. There remained 63 yards between the building and the old Tuckingmill Foundry Office, on which the company did not contemplate erecting any building 'for the present'. Their offer was that they were willing to give sufficient land for a two yard footpath for the whole length of the property, providing the Council would build a wall about 63 yards long enclosing their new boundary, such wall to have similar foundations and width as the one the Company were now building, but to be only eight feet in height from the road line. The Council were dissatisfied with this offer, and decided to meet representatives of the Company. The outcome is not known. However, there is a pavement there at the time of writing (2014) running along the whole northern boundary of the site. At one point it is very narrow as it skirts what was the old Tuckingmill Foundry Offices.

The distinctive factory façade is ranged from here to the corner of Chapel Road and the old Bickford, Smith and Co. Ltd. Offices. Here can be seen the carved initials of the firm, 'BS & Co' on either side of an arched entrance, which forms such a prominent feature of Pendarves Street. One long entire wall of the 'North Lights' jute spinning building borders the pavement at the western end of the site, and it appears that the Council did not build the wall requested.

Figure 53. The "BS" "&Co" initials carved on both sides of the factory entrance.

In 1911/1912 the Tuckingmill factory began to manufacture Bickford-Cordeau-Detonant, (Bickford detonating fuse) which they exhibited in September 1912 at the

Figure 54. Plan drawn up by ICI Explosives (Ltd) 27/05/1952 (CRO Truro)

Royal Cornwall Polytechnic Exhibition.[29] The new method of firing could be used for volley firing or firing single holes. It consisted of using a thin lead filled tube filled with a medium called trinitrotoluene, which in itself was stable, difficult to detonate but highly explosive. (This is better known today by its initials TNT).[30] This lead tube was connected to the bottom most primer cartridge in a borehole. At the other end of the tube a detonator was attached and to this was attached an ordinary time fuse. The detonator and time fuse were outside the bore hole. The time fuse was fired in the usual way, and exploded the detonator, and instantly the detonation was carried through the lead tube to the primer cartridge. This in turn ignited the explosive (gunpowder).

It was claimed by the manufacturers that more work could be extracted from the explosive by this method of firing, owing to the sudden shock given to the bottom cartridge in a bore hole, rather than in ordinary firing. It was also safer as the detonator was outside the hole, and could not be damaged or moved out of place by tamping. Its earliest manufacture was

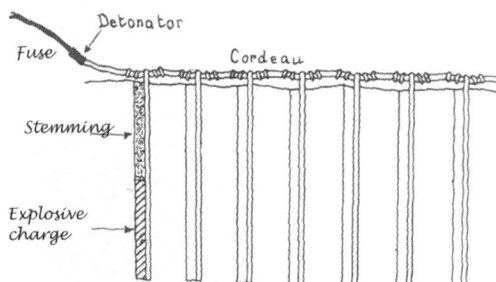

Figure 55. The volley firer.

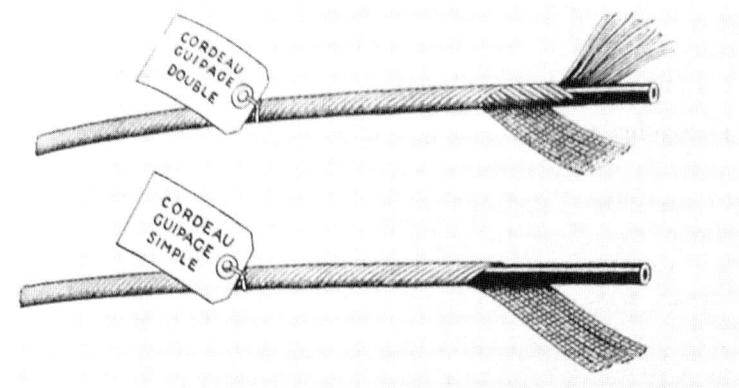

Figure 56. Bickford Cordeau-detonant wrapped with a covering of a single or two cotton threads.

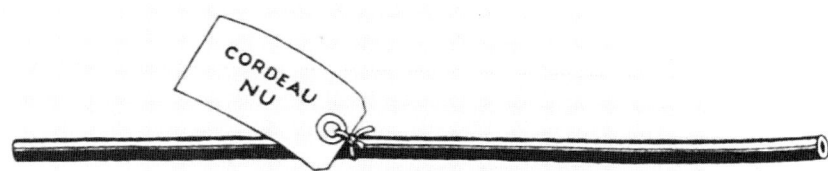

Figure 57. The lead tube is not covered with thread. From 'Manuel Bickford'

in the Davey, Bickford, Smith and Co. Ltd. factory in Rouen, in northern France but it was also made by the Ensign Bickford Company in Connecticut.

In August 1913, a visiting Frenchman, a Mr. Kinsman from a fuse factory in Seyssel, France, gave his impression of the Tuckingmill factory to a local reporter. He noted that Bickford's had 'recently' added a spacious block of buildings for their jute-spinning work. He also noted that Bickford's were manufacturing electric fuses at St. Helen's Junction, Lancashire.[31]

World War I

July 1914 saw the start of the Great War. Bickford Smith and Co. Ltd became heavily involved in producing munitions. Harold Octavius Smith took over as Managing Director of the Company when his brother G. E. Stanley Smith was called upon to serve in his Cornish Regiment.

Some correspondence from the Company with the Ministry of Munitions of War survives in the National Archives, London.[32] David Lloyd George was appointed as the first Minister in May 1915. Large amounts of explosives were being shipped to Tuckingmill. It is not clear whether shipments were by rail, or by boat to a nearby port such as Hayle. Fuses were probably shipped out from Lelant Quay (also known as 'Dynamite Quay') near Hayle.

In October Harold O. Smith wrote to The Director General of Munitions Supply,

Armament Buildings, Whitehall, to inform him that Bickford's had safely received 500 pounds of 'composition exploding' 'which will keep our filling branches occupied until Tuesday evening'. In the same month it was noted that Bickford's windows were to be screened by blinds. It was stated also in October that Bickford's were dealing with 100 pounds per day of 'composition exploding', or trinitrophenyl methylnitramine, known as 'tetryl'.[33]

On October 20th 1915, Bickford Smith and Co. Ltd wrote to Whitehall from their St. Helen's factory saying that the preparations for filling new primers were nearing completion, and that they needed a supply of these primers, together with gunpowder caps.

It seems from the sparse correspondence in the National Archives that Bickford's in Tuckingmill (including Bennett's in Roskear) were filling 'gaines' with tetryl. There is a note in the Archives dated November 3rd 1915, with a heading 'Bickford Smith' reading:-

20,000 Filled Gaines to Armstrong.

A 'gaine' is a short steel metal tube filled with explosive, attached under a nose-cap of a shell and connected to the TNT filling in that high-explosive shell. When the primary detonator is activated, by impact of the shell, the 'gaine' explodes and detonates the main explosive. By acting as a 'booster', it ensures that the detonation of the fuse in the nose-cap of the shell then detonates the contents of the shell. 'Armstrong' referred to the W. G. Armstrong Whitworth and Co. Ltd. munitions factory in Newcastle upon Tyne.

A letter to the Superintendent of the Munitions Stores Department on October 13th 1915 stated that the Raleigh Cycle Company was supplying Bickford Smith and Co. Ltd with empty no. 44 fuses. Bickford's could deal at that time with 6,000 per week, but would shortly 'take on' the filling number again of 10,000. Another note dated November 16th 1915 noted that fuses had been sent to Bickford Smith from the Raleigh Cycle Company, Nottingham. This latter company also turned to making artillery and shell cases during the Second World War.

The No. 44 fuse was introduced

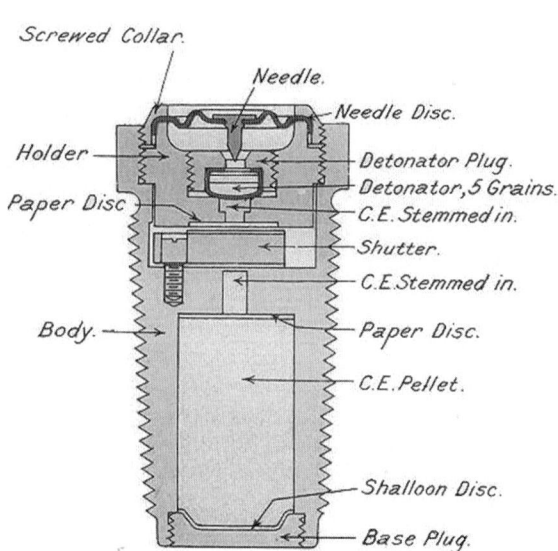

Figure 58. No. 44 Fuze. CE is 'composition exploding' or 'tetryl'.

Figure 59. No. 44 Fuze (about three inches high)

in 1913. A No. 44 fuse or fuze was a small brass cone, truncated at the narrow end, where it was filled with tetryl explosive. The wider end was topped with a copper percussion cap, or disk, with a percussion pin in the middle. When the shell containing the No. 44 fuse was fired from a gun tube (such as a howitzer), the impact pushed the percussion pin against a detonator cap, where a spark fired the detonating charge through a tetryl-filled channel. Tetryl is a nitramine booster explosive. Its function in the No. 44 fuse was to transmit and augment the force and flame from the detonator, in order to propagate the detonation into the main high explosive. It is a yellow, compressed powder material.

The Tuckingmill fuse factories had to put special measures into place to press tetryl into pellets for use in these shell fuzes. The presses were made by Holman Brothers of Camborne.[34] Buildings at the south side of the main Bickford's fuse works seem to have been built to deal with this dangerous material – the 'tetryl' building. The entrance to these buildings was from the western side of the North Lights building on Pendarves Street, and also from Chapel Road.

Clearly the factory was very cramped. In March 1916, Camborne Cricket Club met and agreed to rent their field and cricket pavilion to Bickford Smith and Co. Ltd for one year.[35] The Company probably intended to use the pavilion as a store of some sort (although not for explosives, the storage of which was very tightly regulated). In addition, if explosives were to be stored there, it is unlikely that information about renting to Bickford's would have appeared in the *West Briton* newspaper. During the Second World War, Bickford's used Roscroggan Chapel (about two miles away) as a store also, for timber to make their packing cases. By January 1920, the Cricket Club restarted, with the Roskear grounds totally restored.

Explosives Trades Ltd.

At the start of December 1918, Nobel's Explosives Company Ltd., Glasgow, issued details of the large company formed to combine most of the explosives companies then in operation.[36] The negotiations resulted in a scheme which had the sanction of the Treasury. The shares of the constituent companies would be taken over and held by this new company, to be called 'Explosives Trades Ltd.', and incorporated in England. The new company was going to issue fully-paid shares in exchange for the shares of the existing companies, on the basis of conversion values calculated on uniform lines by two firms of accountants.

The Boards of the following companies passed resolutions in favour of the scheme:-

    Alliance Explosives Company Ltd.
    Australian Explosives and Chemical Company Ltd.
    Bickford, Smith and Co. Ltd.*
    William Bennett, Sons and Co. Ltd.*
    St. Helen's Electric Fuse Co. Ltd.
    Unity Safety Fuse Co. Ltd*
    Birmingham Metals and Munitions Co. Ltd.
    British Explosives Syndicate Ltd.
    British South African Explosives Co. Ltd.
    British Westfalite Ltd.
    Cotton Powder Co. Ltd.
    Curtis and Harvey Ltd.
    E. C. Powder Company Ltd.
    Eley Brothers Ltd.
    Eley Bros (Canada) Ltd.
    Abbey Improved Chilled Shot Co. Ltd.
    Elterwater Gunpowder Co. Ltd.
    King's Norton Metal Co. Ltd.
    Kynoch Ltd.
    Kynoch-Arklow Ltd.
    National Explosives Co. Ltd.
    New Explosives Co. Ltd.
    Nobel's Explosives Co. Ltd.
    Electric Blasting Apparatus Co. Ltd.
    Patent Electric Shot-Firing Co., Rohurite & Ammonia Ltd
    Sedgewick Gunpowder Company Ltd.
    W.H. Wakefield and Co. Ltd.
    *Cornish Companies

The authorised capital was £18,000,000 consisting of £6,500,000 6% Cumulative Preference shares, £8,000,000 Ordinary shares, £1,500,000 Deferred shares and £2,000,000 other shares (the character of which were to be decided at a later date). All the shares were to be of £1 each.

Chairman and Managing Director of the new company was Sir Harry McGowan KBE (Managing Director of Nobel's Explosives Company Ltd). Sir George Smith (Managing Director of Bickford, Smith and Co. Ltd.) was deputy-chairman. There were eighteen directors in all.

The side of the scheme which was of the greatest interest to the shareholders was the terms upon which their shares were to be exchanged for those of the new company. The conversion value of the ordinary shares of the companies was based upon a valuation of the business existing at December 31st, 1915. The goodwill consideration was calculated on pre-war earnings.

One half of the sum for assets was represented by Preference Shares of the new company at par, and the other half and the goodwill by Ordinary Shares at par. The conversion value of all expenditure on capital account incurred for war purposes was to be represented by deferred shares of an amount equivalent to 20% of the cost of the plant and buildings so erected, whiles the growth of assets between December 31st 1915 and December 31st 1916 was to be satisfied by Preference, Ordinary, and Deferred shares.

It was felt that closer co-operation between these companies was necessary to deal effectively with the problems that would arise post-war.[37] Sir George Smith would have very aware of the decline in the Cornish tin mines on his doorstep. Tin prices slumped after the war, and even a great Cornish mine, such as 'The Queen of Cornish Mines', Dolcoath, closed in 1920, as did Wheal Grenville Mine in Troon. The Basset Mines closed in 1919. South Tincroft closed in 1921. Carn Brea Mine had already closed. The employment figures for Bickford, Smith and Co. Ltd tell their own story.[38]

1915	422
1918	378
1919	229

In October 1918, Bickford, Smith and Co. Ltd lost the manager of their munitions works when J. S. Vivian Bickford died suddenly at his lodgings in Camborne, shortly after returning from Mass at the Catholic Church on Beacon Hill. He was the grandson of the inventor of the safety fuse, William Bickford.[39]

After the war, industries such as shipbuilding, coal and steel contracted. Railway workers went on a nine-day strike over pay on 27th September 1919. In July 1919, coal miners in Yorkshire and Wales went on strike. In October 1919, *The Cornishman* reported that both Bickford, Smith and Co. Ltd. and Wm. Bennett, Sons and Co. Ltd. in Roskear had resumed working. Because of a shortage of material owing to the rail strike, they had previously suspended operations.[40]

By November 1920, both Bennett's and Bickford's factories were experiencing difficulty in staying open because of lack of coal caused by another coal strike. Bickford's expected to close 'for a period' while Bennett's informed the local paper that they only had enough coal to last a week or ten days.[41]

Unemployment relief committees had been set up by this time. Their aim was to try and provide work for unemployed miners and also provide small sums of money to help the destitute men and their families. The staff and workers in both Bickford's and Bennett's fuse works contributed to the relief fund. Their total contribution for the third week in March 1921 was £21 16s, bringing their total contributions to £87 10s. A month later, their total contributions had increased to £132 4s 9d. In the same newspaper, the following notice appeared: - 'Sixteen of the few remaining employees of Dolcoath Mine were discharged on Saturday'.

The fortunate employees at Bickford, Smith and Co. Ltd. seemed to have had quite a good social side to their employment. A group from the factory was awarded a First Class Certificate as a reward for their efforts in the Cornwall Folk Dancing Festival in Penzance at the end of June 1920.[42]

Explosive Trades Ltd. becomes Nobel Industries Ltd

The name of Explosive Trades Ltd was changed to Nobel Industries Ltd in October 1920. Explosive Trades Limited issued at that time £3,000,000 8% seven year secured notes at 96½ per cent, with interest payable twice yearly.

The object of the issue was to purchase shares in General Motors Corporation and for the general purpose of the Company. The Company felt the change of name was necessary, because 80% of its capital was used in the manufacture of commodities other than explosives, and it was felt their name had 'become misleading'.[43]

Sir George Smith, the Managing Director of Bickford, Smith and Co. Ltd, died on October 9th 1921, He was buried in his family vault in the cemetery of Camborne Centenary Chapel. The Cornishman reported that about 450 employees of the Tuckingmill and Roskear fuse works marched in procession to the Chapel.[44]

The management of the fuse factory was now in the hands of G. E. Stanley Smith, OBE and his cousins, who were the twin sons of the late William Bickford-Smith.

References, Chapter 9

1. *Bickford, Smith and Co. Ltd. 1831-1931.* The Trevithick Society. 2006
2. *The Cornishman*, March 24, 1900

3. See Reference 1, Chapter 3. G. L. Wilson
4. *The Cornishman*, March 29, 1900
5. *The West Briton*, July 17, 1902. 'The tramway has entered the confine of Camborne Parish. Tuckingmill people are beginning to taste the sweets which have been for some time the fare of the folk at Illogan Highway and Pool.'
6. *The Cornishman*, August 24, 1905
7. *The Cornishman*, January 31, 1901
8. See Reference 1, Chapter 3. G. L. Wilson
9. *The Cornishman*, April 26, 1906
10. Official report of the fire. Fire in Spinning Room at Factory no. 60, Cornwall. Accident No. 106, 1906. Capt. A. P. H. Desborough. HM Stationery Office.
11. *The Cornishman*, June 7, 1906
12. *The Cornishman*, January 31, 1907
13. Maurice Wm. *The Shot Firers Guide*. (Available online). When the sulphuric acid comes into contact with the chlorate of potash, a powerful oxidising agent called chlorine peroxide is produced. When this is generated in the presence of sugar, the vigorous oxidation sets fire to the sugar.
14. Cornwall Record Office documents.
15. See Reference 1, Chapter 3. G. L. Wilson
16. *The West Briton*, January 7, 1907
17. *The Cornishman*, February 14, 1907
18. *The Cornishman*, July 25, 1907
19. *The West Briton*, April 11, 1907
20. *The Cornishman*, April 5, 1908
21. Cornwall Record Office document.
22. *The West Briton*. October 22, 1908
23. See Reference 1, Chapter 3. G. L. Wilson
24. See Reference 1, Chapter 3. G. L. Wilson
25. *The West Briton*, March 28, 1912
26. *The West Briton*, March 28, 1912
27. *The West Briton*, April 11, 1912
28. *The West Briton*, May 6, 1912
29. *The West Briton*, September 5, 1912. The diagram is adapted from 'Manuel Bickford'.
30. Trinitrotoluene is a pale yellow, solid organic compound. Formula: $C_7H_5N_3O_6$. It is relatively insensitive to shock and cannot be exploded without a detonator. It is used in munitions. It is a high explosive but not to be confused with dynamite
31. *The Cornishman*, August 21, 1913
32. National Archives, Kew. MUN 4/1525
33. Tetryl ($C_7H_5N_5O_8$) is commonly used to make detonators and explosive booster charge. Workers who breathed tetryl-laden dust complained of nosebleeds, nausea, vomiting,

coughs, headaches, fatigue and lack of appetite. Workers who handled tetryl developed yellow staining of the hands, neck and hair. Some also developed skin-rashes, and some had asthma-like reactions. Bryan Earl (see below) states that the detonation front passed through the 'Cordeau Bickford Detonant' at from 4,500 to 5,000 metres per second.

34. Earl, Bryan. *Cornish Explosives*. The Trevithick Society. 1976.
35. *The West Briton*, March 23, 1916
36. *Western Mail*, December 2, 1918
37. *Cornish Telegraph*, December 11, 1918
38. Cornwall Record Office. BY/324
39. *Cornish Telegraph*, October 30, 1918
40. *Cornishman*, October 15, 1919
41. *Cornishman*, November 3, 1920
42. *The West Briton*, July 1, 1920
43. *Western Daily Press*, October 26, 1920
44. *Cornishman*, October 19, 1921

Figure 60. Interior of yard building, possibly former blacksmiths' building. 1994
© English Heritage

Figure 61. The derelict North Lights building on Pendarves Street. South Crofty mine behind. 2012. Author's photo.

Figure 62. Bickford, Smith and Co. Ltd. 1924. The buildings at top right centre were built for munitions work during WW1. © English Heritage

Figure 63. Bickford, Smith and Co. Ltd. 1938. The pavilion (1929) (top right centre) can be seen. © English Heritage

Figure 64. Photograph taken in 1964, not long after the fuse works had closed, looking north. © English Heritage

Figure 65. HRH Prince George, Duke of Kent (left) on a visit to Bickford, Smith and Co. Ltd. in May 1932. PB-S Archives

Figure 66. The Sports Pavilion at Bickford, Smith and Co. Ltd., Tuckingmill, shortly after its construction in 1929. PB-S Archives

Figure 67. Foremen and forewomen outside the Sports Pavilion. 1933. PB-S Archives

Figure 68. John C. Bickford-Smith and his twin brother William N. Bickford-Smith. PB-S archive.

CHAPTER 10

Bickford, Smith and Co. Ltd. - The ICI Years and Closure 1922-1961

The factory Directors were now John Clifford Bickford-Smith (1881-1941) and his twin brother William Noel Bickford-Smith (1881-1939). They were the sons of William Bickford-Smith by his second wife Anna Mathilda Bond. Colonel G. E. Stanley Smith was Managing Director. The works manager was C. E. Tyack (a grandson of Dr George Smith), J. W. Lean was sales manager and James Vivian was the engineer.

Nobel Industries Ltd. Local Board
On May 12, 1922, a Local Board was formed for the management of Bickford, Smith and Co. Ltd. Each member was to have the title "Local Director".[1]

Col. G. E. Stanley Smith (Chairman)
Mr. J. C. Bickford-Smith
Mr. W. N. Bickford-Smith
Mr. H. O. Smith

It was agreed that the Seal of the Company could 'be affixed to any document provided that such document be also signed "for and on behalf of Nobel Industries Ltd, Sole Director and Manager" by at least one Director or Local Director, and countersigned by the Secretary, or other officer appointed by the Board'.

Bickford, Smith and Co. Ltd was clearly a loyal employer. The oldest employee, William Henry Fray, died in June 1923, aged 79 years. He had been an engineer with the Company for fifty years.[2]

Nobel Industries Ltd. advertised the sale of

Figure 69. Harold Octavius Smith

Debenture Stock in February 1923. This consisted of £1,750,000 five and a half percent debenture stock at £99 per cent, which would not be repaid earlier than 1928. The terms stipulated that none of the Constituent Companies (which included Bickford, Smith and Co. Ltd) could increase any mortgages or charges on their assets 'except to secure bankers' loans or overdrafts' without the previous consent of the Directors.[3]

In 1923, the Local Board erected a plaque on the factory wall bordering Pendarves Street.[4] It reads: - 'This tablet commemorates William Bickford who invented and first made Safety Fuse at this factory in 1831'. The present tablet is not the original tablet from 1923.

Figure 70. From a photograph in *The Western Morning News*, December 1923. Note the letter 'A' in 'William' and 'factory'.

Figure 71. The present plaque (2012).

Imperial Chemical Industries Ltd.

In December 1926, Imperial Chemical Industries Limited was formed from Nobel Industries Ltd, Brunner Mond and Co. Ltd, The United Alkali Company Ltd, and British Dyestuffs Corporation Ltd. The Chairman was Sir Alfred Mond, M.P., with Sir Harry McGowan as President and Deputy Chairman. Bickford, Smith and Co. was now part of a very large conglomerate.

Figure 72. The Trelawney Pierrots c.1924. Album at CRO

Figure 73. Colonel G. E. Stanley Smith presenting the long service awards at Trevarno, July 12, 1928

Figure 74. A presentation ceremony inside the fuse factory yard. Bickford, Smith and Co. Ltd. 1924. Cornwall Record Office

Figure 75. A presentation ceremony inside the fuse factory yard. Bickford, Smith and Co. Ltd. 1924. Cornwall Record Office

Figure 76. Female fuse factory workers. Bickford, Smith and Co. Ltd. 1924. CRO.

Figure 77. Female fuse factory workers. Bickford, Smith and Co. Ltd. 1924. CRO.

Figure 78. The Welfare Hut 1924

Figure 79. Coilers 1924

The explosive industries of Nobel Industries Ltd. became the Explosives Division of ICI in 1932. In 1948 this Division was renamed the Nobel Division. The remaining interests formed the basis of three other Divisions – Metals, Paints and Leathercloth. Harold O. Smith became Chairman of the Board of ICI Metals Division and was made a Director of ICI in 1936.

The social life of the Bickford's employees was evidently well catered for by the Company. In 1924, the employees were treated to a visit to the British Empire Exhibition in Wembley in London, where Nobel Industries Ltd. had a stand. They were photographed on the steps of St. Paul's Cathedral. A folk dancing group was formed after the War, and gave exhibitions locally.[5] Around this time a Sports Pavilion was built (1929) on open ground to the southwest of the factory, and tennis courts were laid down in front of it. It was a popular venue for dances, whist drives and plays performed by the Dramatic Society.

In July 1928, the annual picnic was held at Trevarno, and long-service awards were presented by Colonel G. E. Stanley Smith. All the employees present can be found in the Appendix 2.

Miss Violet Irene Boot presented prizes to the employees in October 1929, at the second of a series of whist drives and dances held in the new Sports Pavilion. She was the great granddaughter of Dr George Smith.[6]

It is interesting that the Cornish newspapers were beginning to refer (if somewhat inaccurately) to Bickford, Smith and Company Ltd. as part of ICI. In 1929, at the Royal Cornwall Polytechnic Society Exhibition in St Austell, *The Cornishman* referred to them as 'Nobel's Industries (Messrs Bickford Smith and Co.)', when describing their entry of fuses and detonators.[7]

Purchase of Bennett's Fuse Works, Roskear
The sale of the Pendarves Estate, Camborne, comprising buildings, farms and 1,630 acres, took place on Monday June 16th 1930, and lasted for four days.[8] Bickford, Smith and Co. purchased the disused Bennett's fuse works at Roskear for £500, and also their explosive store and two acres of land for £25.[9] Bickford's site at Pendarves Street was in a built-up area, with public roads on two sides, with expansion southwards towards Dolcoath Mine the only option. It seemed that any future expansion was earmarked for the old Bennett's works, which were solidly built and still in a good state of repair.

On 27 September 1931, Charles E. Tyack died aged 69. He was Bickford's works manager and also a local director. He was the grandson of Dr George Smith. The staff and three hundred female operatives attended his funeral at Camborne Wesley Chapel.[10]

Thousands of people lined the streets of Camborne on the 17th of May 1932, when HRH Prince George unveiled the statue to Richard Trevithick, the famous Cornish

Figure 80. The employees of Bickford, Smith and Co. Ltd. pictured on the steps of St. Paul's Ca

engineer. The Prince visited Holman's engineering works, the Camborne School of Mines and also Bickford, Smith and Co. Ltd in Tuckingmill. At the fuse factory the Prince was conducted around by Colonel Stanley Smith, Mr. J. C. Bickford-Smith, Mr. W. N. Bickford-Smith and Mr. Donaldson (manager of the technical department).[11]

The Structure of ICI Explained at the AGM in April 1932

At the fifth AGM of ICI, held in London in April 1932, the Chairman Sir Harry McGowan explained the organisation of the Company.[12] The subsidiary companies were organised in eight groups. Every group contained a number of separate companies, the products of which were cognate to each other. For all purposes of control and administration they treated the companies that formed one group as a unit. The statutory board of each of these companies was the present company, namely ICI, so that there was a uniform legal controlling authority vested in their own board.

He referred to the 'physical rationalisation' demanded for internal economies. He noted that some factories had been sold, some demolished and some held for fresh development. The physical rationalisation was not complete.

Sir Harry outlined how ICI needed to do much more in eliminating disadvantages which arose from the separate existence of different companies. Steps had been taken to reduce the number of these and "Imperial Chemical Industries (Explosives) Ltd." had been registered the previous week. (Bickford, Smith and Co. Ltd was in this new

occasion of their visit to the British and Empire Exhibition, Wembley 1924. PB-S Archives

ICI Company). He noted that Companies which had already been dispensed with had been liquidated after their assets and business had been transferred elsewhere. He stated that he wanted to make quite clear that liquidation of any of their subsidiary companies represented not a tale of disaster, but a mark of progress in the work of financial rationalisation. He predicted that there would be more liquidations, but that care would be taken to see that the goodwill of the company names would be preserved.

It is worth noting here that Bickford, Smith and Co. Ltd. went into voluntary liquidation in 1946, and was closed by ICI in 1961.

In June 1933, John E. Barbary, of Gwennap, who was the assistant manager of Bickford, Smith and Co. Ltd, was awarded an O.B.E. During the war, he was on the War Office staff as a Director of Artillery and later on the staff of the Director General of the Ministry of Munitions.[13]

The Jute Mill

In July 1933, the Royal Cornwall Polytechnic Society visited the Bickford Smith factory. An anonymous writer in *The Cornishman* newspaper described their visit.[14]

> The factory contains the only jute mill in England. The employees are mainly girls. About 400 are employed and they work under ideal conditions. The jute, taken straight off the bales from India, is combed and flattened into bands, passing down a V-shaped face, under a spiral brush, and looking not unlike

Figure 81. Long-service awards at Trevarno. July 1928. PB-S Archives

a waterfall in miniature. It is collected in containers introduced into another machine, and the bands are reduced and compressed. Then they are spun into cord, and onto spools, and then shifted to another machine, and reduced to a smaller gauge, almost like string – only longer.

The writer than described the manufacturing process of a fuse.

The fine cord which is spun in the mill is now spun in the form of a tube; gunpowder, blended with the care of the expert chemist, is fed into it from a rubber tube connected with a hopper above, the stream of powder passing through a tell-tale glass tube which shows the operator if there is a hitch in supply. The powder is encouraged to enter the fuse by a thread of treated cotton, which has the pleasant task of going in with it, to be fired with it later on. This cotton thread runs the whole length of the fuse, right down the core.

The writer continued:-

The visitors then saw a machine which tests the fuse, and is so finely adjusted that it can detect, by the width of the fuse, whether the powder is there or not. We saw the tube coated with bitumistic substance, with gutta-percha (which is made in the firm's own factory in Ponsanooth); we saw the fuse wrapped in textile material, and finally, being given the finishing water-proofing and appearance coats.

Finally, we saw Bickford's fuse burn happily under water, and, with the help of pellets of something pretty useful in the way of explosive, blow to pieces a perfectly good glass and wood tank, in which the experiment was effected. Had some of the explosive been man-size, there would have been no future for the Polytechnic Society.

An interesting case came before Camborne magistrates in May 1935. Bickford, Smith and Co. Ltd brought a case against two Hayle men for poaching on Upton Towans which were owned by the Company. The fuse company kept explosive magazines on the Towans, and were afraid of the risk of fire with poachers about, as lots of matches were left behind on the Towans. Poaching (for rabbits) was ongoing. The two men were fined a shilling each.[15]

The National Explosive Company Ltd. had closed down its high explosive, or Cordite, manufacturing business on Hayle Towans after the war, but Bickford, Smith and Co. Ltd had taken over their gunpowder magazines. These works in the sand-dunes covered nearly 200 acres. Each small building was isolated from another by banks of earthwork and sand rising up to its roof. In the case of the court case against the poachers, it transpired that there had been a caretaker on the Towans for the previous 26 years, namely a Mr. John Pearce.

Figure 82. Photo of Hayle Towans from *The Graphic*, May 18, 1901.

A Motor Accident

Mr. and Mrs J. C. Bickford-Smith of Trevarno suffered a sad loss on May 14th 1936. Their eldest son Jack was killed in a car accident in South Africa.[16] He had been employed there by ICI at their factory at Modderfontein, in the West Rand, one of the centres of gold mining. His younger brother Michael (25) was thrust suddenly into a more active management role at the Tuckingmill factory. He married later that year, in London. His new bride was the daughter of Dr W.H. Coates, a Director of ICI.

In February 1939, W. N. Bickford-Smith, a local director of Bickford Smith and Co. Ltd. died aged 58.[17] He held the rank of captain in the Devon and Cornwall Light Infantry Volunteer Cyclist Company in Camborne, and saw active service in the War in India and Palestine. He was a delegate director of the Explosives Group of ICI.

Very little documentation exists of the fuse Company's manufacturing side during World War II. Lofts at the fuse works were cleared out because of incendiary danger, and the contents were either lost or destroyed.[18] In the National Archives, several files stamped 'Chief Superintendent, Projectile Development Establishment' describe precisely how Bickford fuses were being tested to provide accurate time delay for longer range shells, (such as anti-aircraft), both with atmospheric pressure and reduced pressure. Experiments were carried out on Bickford fuse for 'Burning under controlled pressure' and 'Variation on rate of burning with pressure'.[19]

It is quite clear that the fuses being produced at Tuckingmill had to follow precise and accurate specifications. They were rigorously tested by Government scientists, and were a vital contribution to munitions in World War II.

Bickford, Smith and Co. Ltd, as in the World War I, found that their cramped site lacked storage capacity. Roscroggan Free United Methodist Chapel (about 1½ miles to the north) was brought into use as a storage depot. The Chapel was used to store

Figure 83. The mess hall and yard c.1930.

Figure 84. Roscroggan Chapel. Date unknown. PB-S Archives

timber, and possibly jute. *The Cornishman* reported on July 12th 1941 that 3 boys, all evacuees, had broken into the Chapel through a boarded-up window, and had stolen a quantity of box boards. The timber was the property of Bickford, Smith and Co. Ltd and was valued at ten shillings.[20]

A much sadder chapter in the history of this small chapel occurred two weeks after this. Minutes after taking off from RAF Portreath at 8.50 am on July 26 1941, a Bristol Beaufort aircraft crashed into the Chapel, with the loss of life of all four Allied airmen aboard.

October 1941 saw the death of John Clifford Bickford-Smith of Trevarno.[21] He was a Governor of Camborne School of Mines and High Sheriff of Cornwall. The works manager at this time was John E. Barbary.

ICI's Role in War-Time Munitions
At the 18th Ordinary General Meeting of ICI in London on 31st May 1945, Lord McGowan outlined ICI's war-time role in manufacturing munitions.[22]

> Since 1939 almost 400,000 tons of explosives have been made, besides hundreds of millions of detonators, fuses, and other accessories, including the filling for over 90 million incendiary bombs.
>
> Throughout the war our Explosives Division rendered invaluable service in the realms of research, experiment and development. One of the notable developments of the war has been the triumph of explosives over steel and concrete. This came about as a direct result of applying the practice of modern

Figure 85. Miss Salome Lawrey (centre) and fellow fuse workers at Bickford's. 1928. Courtesy Roy Blewett.

Figure 86. Miss Winnie Simmons at Bickford's in 1951. Courtesy Roy Blewett.

quarry-blasting to the uses of war. This was done on a number of devices, using plastic explosive, in the development of which your company played an important part, besides producing them in great numbers.

Lord McGowan then spoke about the delay action shells (which contained Bickford fuses):- 'Our experts devised a range of delay action detonators for aircraft bombs which was so much better than previous types that it was standardised by the RAF. Practically the whole of the present range of aircraft bomb delay-action detonators is a contribution of ICI.'

Lord McGowan then talked about the labour force during the war. He spoke of 'the virtually complete freedom from labour disputes'. He reported:- 'this freedom is the more remarkable when it is remembered that the employees who have been brought up in the tradition of the Company have been heavily diluted by new entrants directed in many cases from distant parts of the country who know nothing about ICI or the aims and objects for which it exists'.

In June 1945, James W. McDavid, Managing (delegate) Director of the Explosives Division, Imperial Chemical Industries (Explosives) Ltd was awarded a CBE in the King's "Victory" Honours List.

Voluntary Liquidation

In 1942 the policy decision was taken by ICI to liquidate the main Group companies, and turn them into Divisions of ICI Ltd. This was not fully accomplished until 1946. According to Dr. William Coates, the move would simplify and clarify the whole financial and legal procedures. All the employees would become 'the servants of one company' and the move would reduce secrecy amongst the subsidiary companies.[23]

In April 1946 the Limited Company that was Bickford Smith and Co. went into voluntary liquidation, along with British Westfalite Ltd, which was

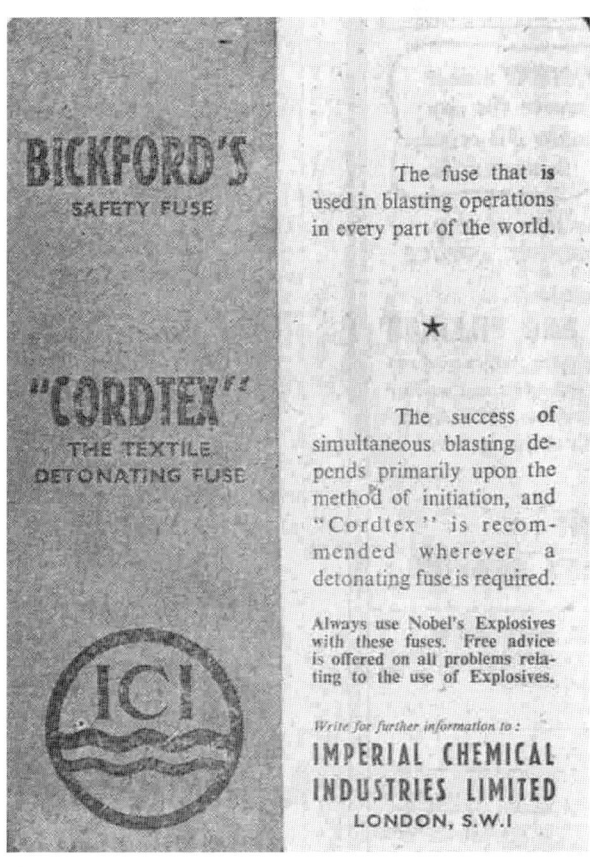

Figure 87. Advertisement from *The Cornishman*, 13th June 1946

Figure 88. Michael Bickford-Smith PB-S Archive.

another member of the Explosives Division of ICI. A general meeting was held at Nobel House, Buckingham Gate, London, on 23rd May 1946 to wind up the Company.[24] Bickford, Smith and Co. was now an integral part of Imperial Chemical Industries, and the Tuckingmill manager now reported to the Head of the Explosives Division.

In August 1946, long-service awards were presented in Tyack's hotel, Camborne, and the employees met with Dr W. H. Coates, a Director of ICI, and with J. F. Lambert, Personnel Director of the Explosives Division. Michael Bickford-Smith (34) was the works manager, having taken over from J. E. Barbary who had died in February 1945. M. M. Forbes, the works superintendent, and J. A. Willacy, the chief clerk, were also present.[25]

The post-war coal shortage was felt keenly in the area, and consumers were urged to cut consumption. The severe winter of 1946/47 buried the UK under large amounts of snow. Coal supplies, already low, could not be delivered, hitting industry particularly hard. Percy M. Holman, of Holman Brothers, Camborne, stated that they were using reserve supplies of steel, while Michael Bickford-Smith reported that work was continuing 'at full speed' for the time being.[26]

At the start of September 1947, Camborne and Redruth Fire Brigades were called to a heath fire close to South Crofty Mine. The Bickford, Smith and Co. explosives

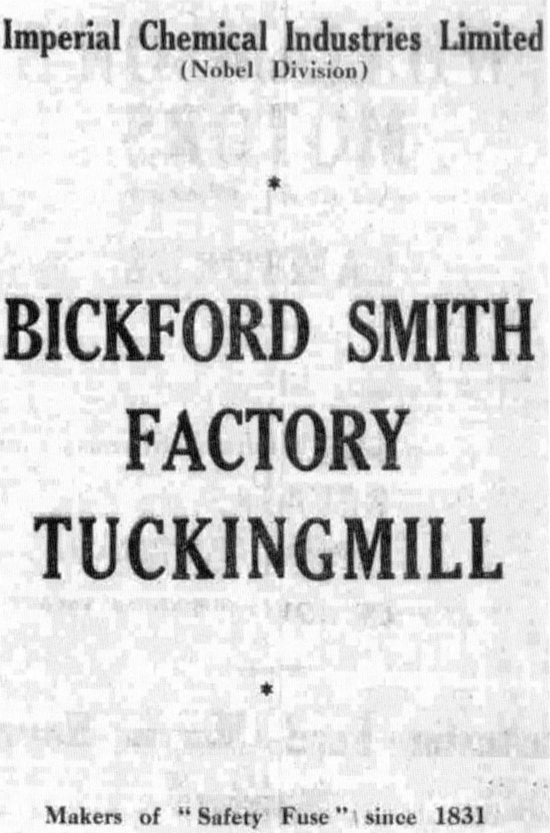

Figure 89. Advertisement from *The Cornishman* 31st August 1950.

magazines were full at the time, and posed a serious threat.[27] Black vans from the Tuckingmill factory were often seen travelling through Pool and turning to go up Druid's Road, in the direction of Carn Brea, from where explosive material was either transferred to, or taken from, the sturdy magazines.[28]

In July 1948, ICI (Explosives) Ltd of Tuckingmill was fined £100 and the deputy manager Goronwy Davies was fined £25 for laying 800 square yards of concrete paths on the sports ground without a licence. In the course of the court case, it was stated that the sports pavilion, to which the roads had been connected, had been used for war industry, but it was now used again for recreational activities. The 'war industry' was not specified.[29]

In 1948, the Explosives Division was renamed the Nobel Division, as its products extended beyond the explosives field, and included many products which, although allied to explosive manufacture, were not explosives. The chief business of the Division was the manufacture of commercial explosives for mining, quarrying etc. and other items such as safety fuse and detonators. The Headquarters of the Nobel Division was in Glasgow. The Division employed 2,000 staff and 10,000 workpeople in 1950. After the war, Nobel was supplying most of India's and South Africa's requirements. However, local manufacture began to steadily grow, putting Nobel into decline. By 1952, Nobel's percentage of the total ICI capital employed in the United Kingdom had fallen from 23% (at the time of the merger) to just 9%.[30]

At the 34th Annual General Meeting of ICI, held on 18 May 1961, the Annual Report for 1960 was discussed. The Chairman was Paul Chambers, who succeeded Sir William H. Coates as Financial Director of ICI in 1948. The Groups and their Directors were listed.[31]

GROUPS	DIRECTORS
Group A.	
Alkali Division	
General Chemicals Division	J. S. Gourlay
Group B	
Dyestuffs Division	
Paints Division	
Pharmaceuticals Division	G. K. Hampshire
Group C	
Fibres Division	
Heavy Organic Chemicals Division	C. Paine
Group D	
Billingham Division	

Nobel Division
Wilton Works & Severnside Works   R. A. Banks

Group E
Metals Division   J. Taylor

The Technical Director of ICI was Dr Richard Beeching, whom the Board of ICI agreed to release to take up his appointment as Chairman of the British Transport Commission, and later of the new British Railways Board from June 1961.

An outline of overseas manufacturing activities for Africa was given: - 'African Explosives and Chemical Industries Ltd (AE & CI Ltd) – the ordinary capital of which is owned in equal shares by ICI (S. Africa Ltd) and De Beers Industrial Corporation Ltd – again increased its sales. Facilities for the manufacture of safety fuse should be ready by the middle of 1961'.

The safety fuse factory was located at Modderfontein, near Johannesburg. An explosives factory had first been opened there in 1896 to produce explosives for the gold mines. The production of safety fuse by AE & CI Ltd commenced in 1961.

AE & CI Ltd had taken all of Bickford Smith and Co.'s output up until then. Because production was moving to S. Africa, the imminent closure of the Tuckingmill factory was announced at the end of 1960. The local Member of Parliament, Frank Hayman (Lab), raised the matter of the closure in the House of Commons.

ICI closed the fuse factory of Bickford Smith and Co at the end of July 1961, thereby bringing to an end 130 years of fuse production on the site. The lease for the fuse factory site expired on the last day of December 1961. Parts of the fuse factory buildings are today occupied by small business concerns and light industry. Some of the buildings on the south side have been demolished, while the iconic North Lights building is derelict and has been for some years. English Heritage has listed as 'Grade II' the Front Range to the former Bickford Smith's fuse works.

References, Chapter 10

1. CRO document
2. *The West Briton*, June 21, 1923
3. *Aberdeen Press and Journal*, February 12, 1923
4. *The Western Morning News*, December 22, 1923
5. *The Western Morning News*, June 23 1927
6. *The Western Morning News*, October 24, 1929
7. *The Cornishman*, February 21, 1929
8. *The Cornishman*, June 19, 1930

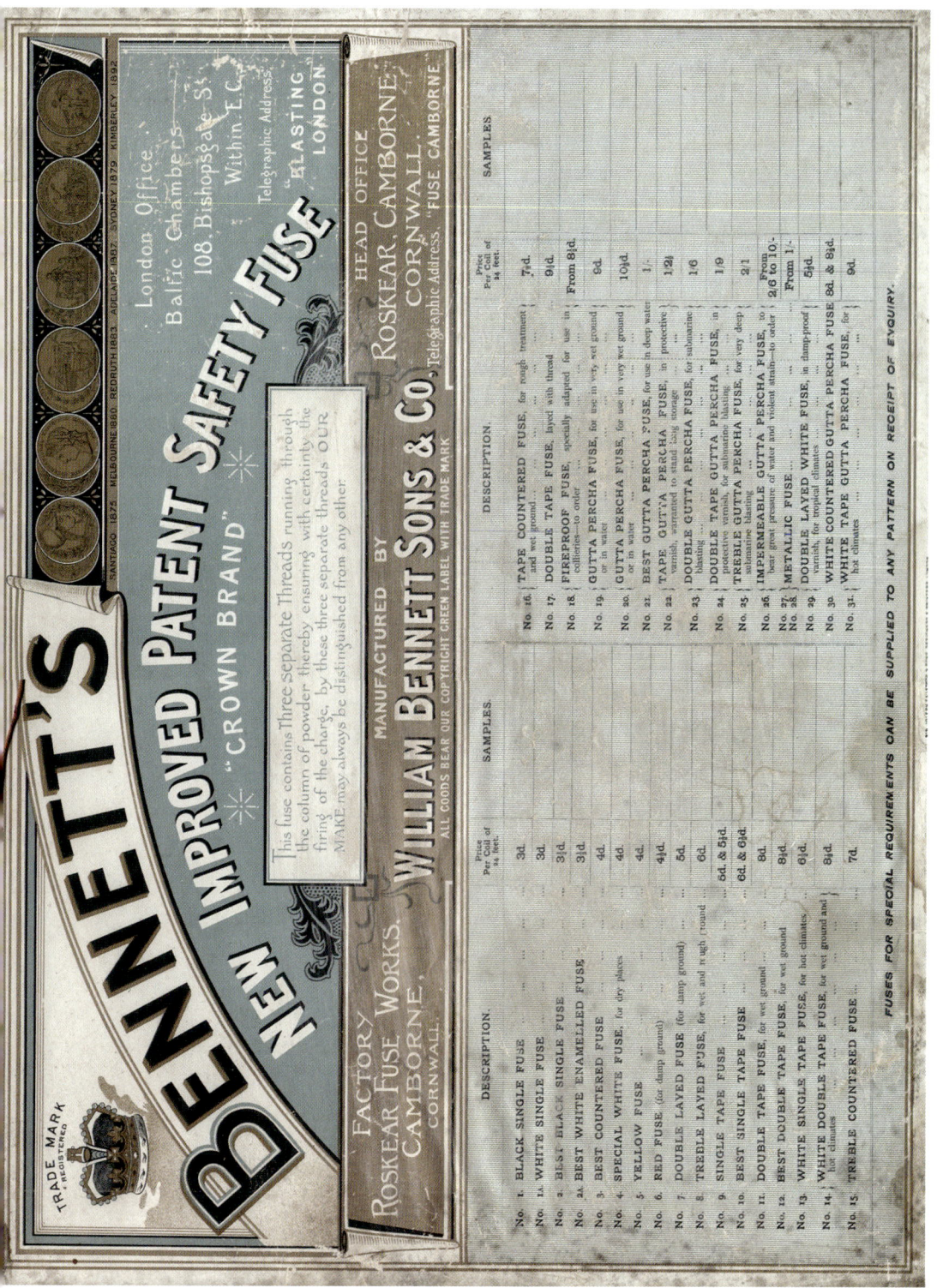

Figure 90. Bennett & Co. price list, date unknown. Trevithick Society.

Figure 91. Fuse colour samples from Bennett & Co. Trevithick Society.

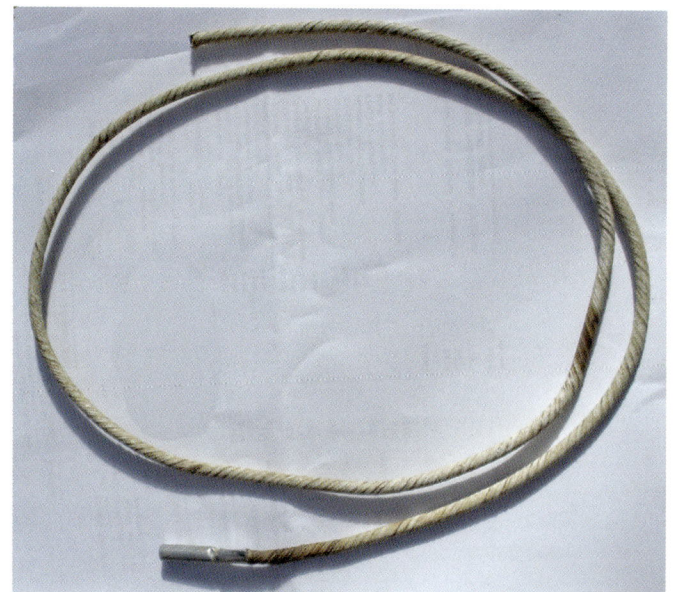

Figure 92. This Bickford fuse is WCGP (White Countered Gutta Percha). The detonator is a No. 6 aluminus ASA (initiating charge – Azide Styphnate) PETN (base charge – Pentaerythritol Tetanitrate). It is a very late example of the fuse manufactured at Tuckingmill.

Figure 93. Cordtex. Bryan Earl is holding a live open-cast gelignite cartridge primed with Cordtex fuse.

Figure 94. The Bendigo Fuse Factory, Wattle Street, Victoria, Australia. Heritage Victoria.

Figure 95. The Institute, Porthleven.

Figure 96. Rear of North Lights building. The original whitewash can just be glimpsed. 2013

Figure 97. Chapel Road, 2013.

Figure 98. The Sports Pavilion 2013

Figure 99. Fuse factory façade looking east. 2013.

Figure 100. The Bickford, Smith & Co. Ltd. premises and Pendarves Street, Tuckingmill. 2013

Figure 101. North Lights building. Pendarves Street. 2013

Figure 102. These buildings were used for making safety fuse and metallic fuse. 1994.
©English Heritage

Figure 103. The buildings where fuses were made are now apartments. 2012.

Figure 104. Yard building in 1994. © English Heritage

Figure 105. The same building in light industrial use. South Crofty mine behind. 2012.

129

Figure 106. Derelict building at rear of North Lights, use unknown. South Crofty Mine headframe at rear. 2013.

Figure 107. The east and west entrances to the fuse factory. 2013.

Figure 108. Former Bennett & Co. offices at Roskear in 2013. Peter Joseph.

Figure 109. Rear of offices, 2012; This end was used as an electricity substation by SWEB. Author's photo.

Figure 110. Bennett's Fuse Works on Roskear Terrace, 2013. Author's photo.

Figure 111. Cut stone entrance to Bennett's Fuse Works, 2012. Author's photo.

Figure 112. Bennett's Fuse Works, 2012. Author's photo.

Figure 113. Rear of Bennett's Fuse Works, 2012, the former Norman's Cash & Carry. Author's photo.

Figure 114. Tremar Combe, the Tremar fuse works looking south. 2014.

Figure 115. Tremar Combe, the Tremar fuse works looking north. The Chapel roof can be seen.

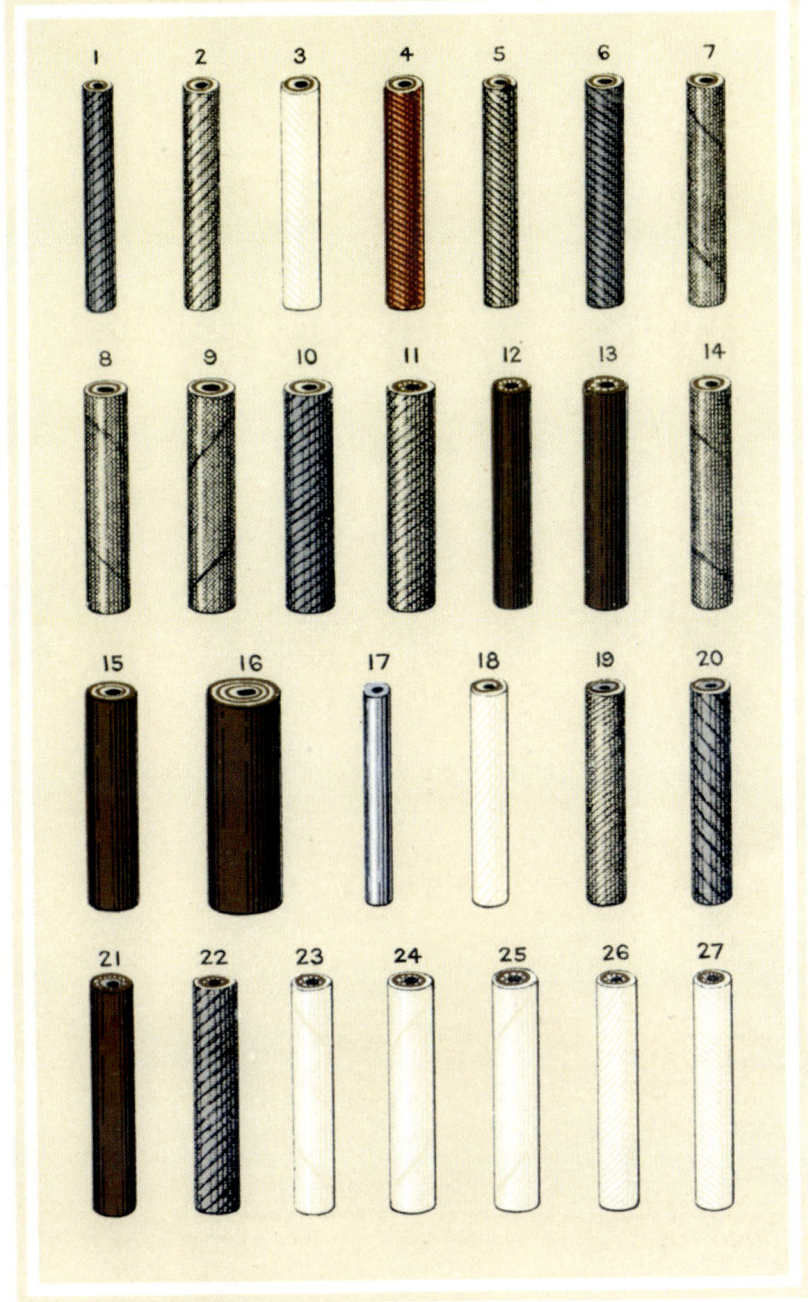

Figure 116. Examples of fuse colours from the Bickford, Smith & Co. Ltd. catalogue, 1895.

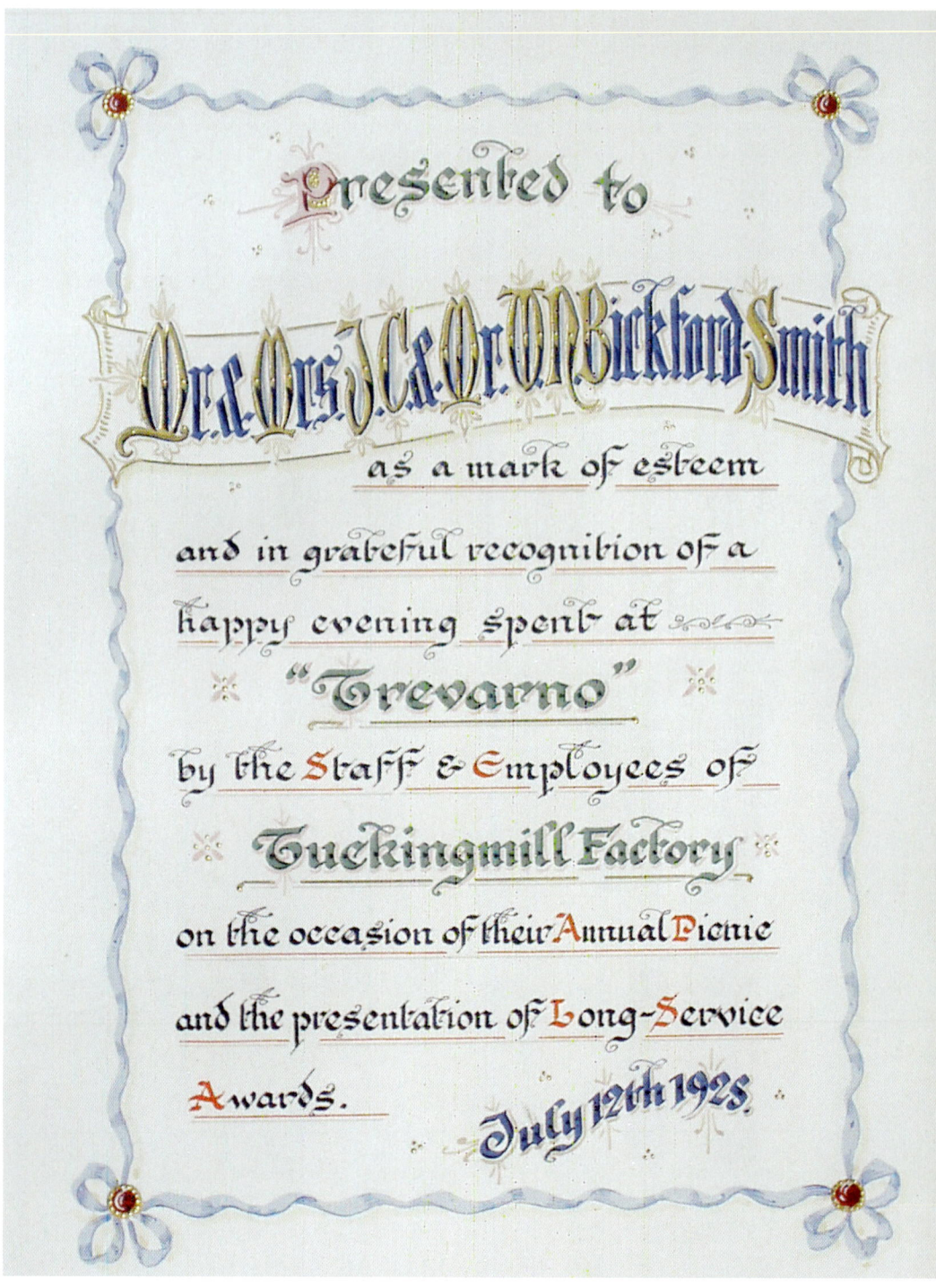

Figure 117. Front page of the long-service awards book. PB-S Archive.

9. *The Cornishman*, June 23, 1930
10. *The Cornishman*, October 8, 1931
11. *The Western Morning News*, May 18, 1932
12. *The Courier and Advertiser*, April 15 1932
13. *The Cornishman*, June 8, 1933
14. *The Cornishman*, July 27, 1933
15. *The Cornishman*, May 23, 1935
16. *The Cornishman*, May 21, 1936
17. *The Cornishman*, February 23, 1939
18. G. L. Wilson. See reference 1, Chapter 3.
19. Files at the National Archives, Kew. AVIA41/27 and AVIA/27
20. *The Cornishman*, June 12, 1941
21. *The Cornishman*, October 9, 1941
22. *The Cornishman*, May 31, 1945
23. Pearcy, Jeff. *Recording an Empire. An Accounting History of ICI. 1926-1976.* Published by the Institute of Chartered Accountants of Scotland.
24. *The London Gazette*, April 12, 1946
25. *The Cornishman*, August 1, 1946
26. *The Cornishman*, February 20, 1947
27. *The Cornishman*, September 4, 1947
28. Personal communication by a local resident to the author.
29. *The Cornishman*, July 8, 1948
30. Pearcy, Jeff, *op. cit.*
31. ICI Annual Report for the Year 1960.
32. G. L. Wilson. See reference 1, Chapter 3.

PART 2

THE CAMBORNE AREA FUSE WORKS

CHAPTER 11

William Brunton's Fuse Factory, Penhellick, Pool
1846-1898

William Brunton was the son of a famous Scottish engineer, William Brunton senior (1777-1851). When he was a young man, William senior worked at Boulton and Watt's Soho Manufactory in Birmingham. He worked closely with James Watt, and became superintendent of the engine manufactory. From 1824 to 1826 he was the engineer for the Redruth and Chasewater Railway, which ran from Point Quay at Devoran to Redruth. He moved to London to open an independent practice as a civil engineer and in 1828 took out a patent for his well-known calciner. This was used in Cornwall to extract arsenic from arsenic-rich tin ores. After a most distinguished career, much of it spent in Wales, he retired to Cornwall.

In 1835, William Brunton junior, also an engineer, emigrated to the United States of America, where he became the Locomotive Superintendent of the New Orleans and Pontchartrain Railway. He returned to the North of England in 1839, where he worked on the Manchester and Leeds Railway, which opened in 1841. He then went back to the United States, where he worked on the Red River Canal.[1] He crossed the Atlantic once more, and came to reside in Cornwall in 1846, when he was appointed resident engineer of the West Cornwall Railway. His son Charles Robert was born on July 31st 1847, when he and his wife Jane were residing with their two other children in Pool, near Camborne. By 1851, he was living in Camborne Cross with his family and his father William (74). Both were listed in the census as 'civil engineers'.

The chief engineer for the West Cornwall Railway was Isambard Kingdom Brunel. The company was formed in 1846 to construct a railway between Truro and Penzance, taking over the Hayle Railway of 1834, which linked Hayle and Redruth. In March 1852, the line was opened between Penzance and Redruth, and was eventually extended to Truro in 1855. The West Cornwall Railway was sold to the Associated Companies in 1865-66. This was a group comprised of the Great Western Railway, the Bristol and Exeter Railway and the South Devon Railway. In taking over the Hayle Railway, however, the West Cornwall Railway also took over their workshops at Carn Brea, just south of Pool. These workshops became the new Company's engine repair and manufacturing centre. William Brunton junior opened his fuse factory very close to these works while he was still working as engineer for the railway.

The original fuse factory was situated east of the road which led from Pool Village to Tincroft Mine. The old Penhellick copper mine formed a part of this ancient mine, which had been in operation since the 18th century at least. During the period when Brunton's fuseworks commenced (1845-46), copper was being mined in sufficient quantity to enable dividends of £24,000 to be paid to the Tincroft investors from 1843 to 1846.[2] These fuse works were established so quickly after Brunton arrived in Cornwall that it is possible that disused mine buildings were occupied. The long low buildings that can be seen in the 1924 aerial photograph are probably the buildings erected after the 1865 explosion that destroyed the works. Immediately to the east was a productive copper (and later tin) mine, East Pool, which reopened in October 1834.[3] South of the fuse factory were the Carn Brea mines, comprised of four old copper mines: - Tregajorran, Wheal Fanny, Wheal Druid and Barncoose. They amalgamated in 1832 to form Carn Brea Mines.

Brunton may have had assistance to start the factory. It transpired in 1862 that one of his partners was Walter Pike. His father, Robert Hart Pike, was the Secretary to

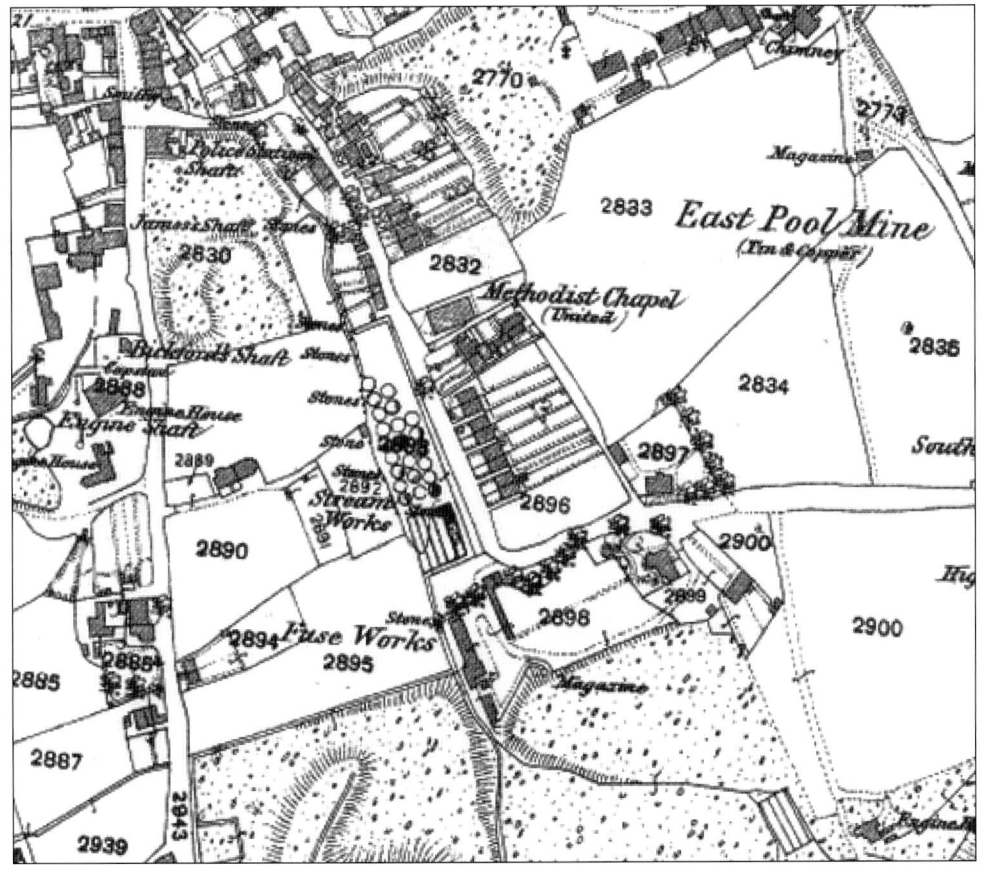

Figure 118. Brunton's Fuse Works on the 1880 Ordnance Survey map.

Hayle Railway Company in 1839, in their London Office at 85 London Wall. Next door, at no. 86, was the office of the Carn Brea Mining Company. R. H. Pike came to Cornwall in the 1840s as Superintendent of the West Cornwall Railway. He was Secretary/Purser to the Carn Brea Mines when he was living in Cornwall. He may, however, have been the Secretary while still living in London. On the birth certificate of his three London-born children, he described himself as 'Secretary' (1836) or 'Secretary to a Mining Company' (1838 and 1840).[4] As an influential figure in these successful mines, he would have been in a position to choose the manufacturers who supplied the mines with safety fuse. He and his son Walter eventually became pursers to many of the Camborne and Redruth mines.

In this venture, William Brunton had the help of two young mechanical geniuses, namely James and Joseph Tangye. With their three brothers, they eventually founded the Cornwall Works in Birmingham. This world famous engineering manufactory had 2,000 employees by 1880. James Tangye (1825-1913) first worked at Copperhouse Foundry, Hayle. He then worked in the Redruth Foundry, and then in an engineering firm in Devonport. He then returned to his native village, Illogan, and joined William Brunton (junior) in making machinery for the manufacture of safety fuse. His brother Joseph was already working there and they both remained there for a number of years. In his autobiography, Richard Tangye (1833-1906) noted that his brothers' machinery greatly lessened the cost of fuse production.

The factory was manufacturing fuses by 1849. In November of that year, a Glasgow firm of merchants by the name of 'Long and McLean' was placing advertisements in the local newspaper stating they had been appointed sole agents for the sale of Messrs. Wm. Brunton and Co.'s 'Patent Safety Fuse'.[5]

An interesting case came before the Cornwall Midsummer Sessions in July 1850. James Chatten, a civil engineer from Tuckingmill, applied for a licence to build a gunpowder magazine at Upton Towans in Gwithian. He stated that it was needed in order to store gunpowder from manufacturers outside Cornwall. He described how William Brunton could not get gunpowder for his safety fuses from Messrs. Sampson and Co. Chatten told the court that this seemed to imply that Sampson and Co. was unable to fully supply all their customers, and that a gunpowder magazine would enable gunpowder landed at Hayle to be stored in barrels. The application was refused because of objections.[6]

Brunton's factory claimed to have supplied the Polar Expedition of Sir John

SAFETY FUSE.—Messrs. WILLIAM BRUNTON and CO., PENHALLICK, near REDRUTH, CORNWALL, MANUFACTURERS OF FUSE, of every size and length, as exhibited in the Great Exhibition of 1851, and supplied to the Royal Arsenal at Woolwich, the Arctic Expedition, and every part of the globe.
Messrs. BRUNTON & CO. are at all times PREPARED to EXECUTE UNLIMITED ORDERS for SUPPLYING FUSE direct from their own MANUFACTORY, upon warrant that it will prove equal to, if not better, than any to be procured elsewhere.

Figure 119. Advertisement from *The Mining Journal*, 3rd December 1853.

Franklin with waterproof fuses covered in gutta percha. The Expedition sailed in May 1845. The Company exhibited its fuses in the Great Exhibition held in the Crystal Palace, London in 1851.They did not win a medal however; this honour went to their Tuckingmill rivals, Bickford, Smith and Davey. In the same Exhibition, William Brunton junior entered a 'machine for washing ores', known as 'Brunton's Endless Cloth'.

The West Cornwall Railway
In March 1852, Brunton showed the Directors of the West Cornwall Railway around the Carn Brea Engine Works. The line from Redruth to Penzance was in the process of being opened, and the Directors decided to take their first trip on this line. On the last Saturday in February they assembled at Redruth. They were accompanied by Robert Hart Pike, the Superintendent of the railway and by William Brunton.[7]

> The train consisted of five carriages, one of which was a very handsome and convenient – we might even say luxurious – first-class, in three compartments, built at the Company's manufactory at Carn Brea. The carriage excited much surprise and admiration, by the excellence of its workmanship, its well-planned arrangements for lighting, ventilation, and general convenience, and the elegance of its ornaments. The engine employed on this occasion was appropriately named the *Penzance*. We believe it was built at Carn Brea, where, under the direction of Mr Brunton, superintending engineer, another powerful engine is in the course of making – to be called the *Camborne*. This will be followed by a third, to be named the *Redruth*, and a fourth, to be called the *Truro*, is expected to be completed by the time (probably in July or August) when the line from Redruth to Truro will be completed.
>
> On the arrival of the train at Carn Brea, the Directors got out to examine the works of the new engine and carriages, and other plant in process of manufactory; and, assisted by Mr. Brunton, very courteously showed the works to their friends and visitors….In proof of the economical arrangements effected at the Carn Brea foundry, we were informed that the Company can build their own engines at the same cost for three which they would otherwise have to pay for two.[8]

One of the Directors of the West Cornwall Railway was George Smith, whose own fuseworks, Bickford, Smith and Davey, was within walking distance of Brunton's factory. Brunton, however, appeared to be fully occupied with the Carn Brea Engine Works; it may have been around this time that James and Joseph Tangye were made partners in the concern.

In December 1852 William Brunton brought a curious case before the County Court at Redruth. He brought an action against a cattle dealer called Trestrail, to

Figure 120. The steamer *Indus*

recover damages for breach of warranty on the sale of a cow. Trestrail had told Brunton that the cow would yield two pounds of butter per day. Brunton brought the case because the cow did not produce enough milk to do this. The judge found in favour of Brunton.[9]

In July 1853, Robert Hart Pike resigned from his position as Superintendent of the West Cornwall Railway. At a presentation of a silver salver at Matthew's Hotel, Camborne, the Chair was taken by William Brunton.[10] Pike became Purser and Secretary to many Camborne and Redruth mines. He continued to live at Trevu Road, near Camborne railway station.

The Tangye brothers left the fuse works sometime in 1854 to join their brother Richard in Birmingham. The factory was not sold, however, and the partners installed a manager called Oliver May and a young clerk by the name of Peter Alfred Renfree from Redruth.[11]

In November 1855, William Brunton was called as a witness in a case where Samuel Vigurs, who was living near the railway line in Besore, near Truro, claimed sparks from the *Camborne* set fire to his house, which then burned down. Brunton was described as 'resident engineer for the West Cornwall Railway when the line was constructed; and continued so till very recently.' The jury awarded damages to Mr. Vigurs.[12]

William Brunton leaves Cornwall

William Brunton had not only left the West Cornwall Railway Company, but was about to leave England. He had been made a Member of the Institution of Civil Engineers by this time.[13] He departed from Southampton, at the start of September 1856, on the steamer *Indus* for Bombay and Karachi. The Directors of the Scinde Railway Company engaged him to survey the country between Moolton, Lahore and Umritsir in the Punjab. Brunton was appointed by the Scinde Company under the sanction of the East India Company.[14]

In January 1856, a local auctioneer, John Burger, put a small notice in the Royal Cornwall Gazette stating that he was going to sell by auction 'the MODERN HOUSEHOLD FURNITURE, COBOURG etc., the property of Mr. James Tangye and Brothers, at Penhellick Safety-fuse factory, near Pool Railway Station'. This implies that one or more of the brothers had been living in an adjoining house to the fuse works.[15]

W. BRUNTON and Co. have great pleasure in informing their customers and friends, and the Mining community, that they have resumed manufacturing at their PENHELLICK WORKS, *Pool*, near Camborne, and are prepared as before to supply SAFETY FUSE of a quality which cannot be surpassed.

Branch Works, Brymbo, near Wrexham.

Figure 121. *Royal Cornwall Gazette.* 31st May 1861.

In July 1860, the partnership that existed between the five Tangye brothers and William Brunton and Co. was dissolved. The remaining partners were William Brunton, Walter Pike, Peter Alfred Renfree and Susan Daniell.[16] The latter was the widow of copper mine agent Francis Daniell. He was Purser of Camborne Consols mine, and had died in April 1860. His widow lived at Polstrong House, which Francis Daniell had purchased in early 1857.[17]

Wednesday April 10th 1861 – A Fatal Explosion

Two young women and a fourteen year old boy were killed when an explosion happened in the fuse-room of the fuse factory. The accident happened at about one o'clock, when most of the workpeople were having their midday meal. The women were Ann Hancock and Elizabeth Blight. The young boy was William Sleeman.[18]

The inquest was held the following day at the *Railway Inn*, Illogan, before a jury of twelve men. The jury inspected the fuse works, and found the premises 'in an extremely ruinous appearance'. They found that the wall of the fuse room at the western end of the building had been blown away. Another wall had been carried across the road that passed by leading up to the railway. The roof of the house had been blown into an adjoining garden, and the timbers were charred by the fire.

The adjoining house where the manager, Peter Renfree, lived, suffered extensive damage.[19]

The jury were then shown the remains of the victims.

Their remains were packed in something like coarse sacks. On opening the mouth of the first of these was seen a fearful object, which was said to be the remains of Ann Hancock. The other poor creature, Elizabeth Blight, was so black that nothing could be seen but her teeth. In the house upstairs, lay in a coffin, the body of the deceased William Sleeman. He appeared not to be injured by fire as he was sitting in an adjoining room.

James Gilbert, an engineer employed by the fuse works gave evidence.

An explosion took part in the western part of the building in the fuse room. The building consists of two floors only, the ground floor and the floor above. The fuse room was upstairs; not quite at the extremity of the building; there was one small room beyond it. The division between the fuse room and the further small room was a dead stone wall. There was winding machinery in a room a long way from the fuse room, and the engine was in another room outside, at the eastern extremity. The whole length of the building is about 100 feet. The machinery was worked with steam. The boiler and the fire-place are without the building. The employment of Ann Hancock and Elizabeth Blight was that of fuse making. At the time of the accident they were in the fuse room; I am not aware that any other persons were there at the time. The deceased, William Sleeman was supposed to be in an adjoining room, the winding room to the eastward. The fuse room had eight machines for making the fuse, and the deceased women were employed in working at them. Sleeman had no business upstairs unless cleaning the benches. If he had been there, he would have been talking to the girls. His general employment was to drive the engine. I cannot account for the accident in any way. My supposition is that some part of the machinery connected with the tubes was broken and that must have caused ignition.

Peter Alfred Renfree (who was twenty-five years old) testified that he was employed as 'clerk and manager'. He stated that he had worked at the fuse factory for between six and seven years. He did not mention that he was a partner in the firm.

Thomas Gilbert, who gave evidence with his head bandaged, said he was in the winding room where Sleeman was, and he was found about eight feet away from him.

Neither of the Gilbert brothers informed the judge that Sleeman was their nephew and that he was only fourteen years old. The jury returned a verdict of 'Accidental

Death'.

The newspaper reported that the works were likely to open again in two or three weeks' time. 'The value of the fuse destroyed is about £10, and the total loss is between £200 and £300. Messrs Brunton and Co. have other works at Wrexham in North Wales, so their business will not be affected by the accident'.

The three people who were killed in the explosion were buried in Illogan churchyard. The *Royal Cornwall Gazette* noted that between 4,000 and 5,000 people were present at the funeral. 'As Saturday is a miner's holiday, most of the vast concourse belonged to that class of persons.'

The census for 1861 was only taken three days before. Ann Hancock (28) was living with her widowed mother in nearby Tregajorran, and gave her occupation as 'safety fuse maker'. Another sister, Jane (22) also worked in the fuse factory. Her four brothers were copper miners. Elizabeth Blight (28) lived in Brea with her parents Jane (63) and Joseph (66), who was a shoemaker. Her brother James (26) was a tin miner. William Sleeman (14) was living in Tincroft Lane with his widowed grandmother Elizabeth Gilbert (63), a milliner, her son Thomas Gilbert (27), a safety fuse maker, and another son James Gilbert (25) an 'Engine Smith'. William Sleeman's occupation was given as 'blacksmith'.

The *Royal Cornwall Gazette* had referred to Brunton and Co's other fuse works near Wrexham, in North Wales. This fuse factory, in Brymbo had been opened sometime towards the end of 1858. An Illogan man, Oliver May, was sent there to open these works, and he continued there until his death in 1902. As William Brunton was in India by now, this move was instigated by the Tangye brothers. Brymbo was the location of an iron works, a blast furnace, coal mines and (later) a steelworks. The fuse works specialised in electric fuses. It was owned after 1860 by Peter A. Renfree, who owned and managed the Penhellick Fuse Works. By 1910 his two sons Alfred and Herbert were managing the Brymbo works, which were also known as the Cambrian Fuse Factory.[20]

PATENT SAFETY FUSE WORKS, BRYMBO, NEAR WREXHAM. – Messrs. BRUNTON & CO., of the Penhellick Fuse Works, Cornwall, having erected a branch manufactory at the above address, beg to inform Merchants and Buyers of Safety Fuse that they are prepared to execute orders to any extent from there, thus obviating the great delay arising from having to send to Cornwall, there being no other manufactory of fuse out of that county.

Figure 122. *Manchester Weekly Times*, **February 19, 1859**

Thursday April 27th 1865 – Another Fatal Explosion
The Royal Cornwall Gazette carried the sad headline: - 'The Fatal Explosion at Pool', and gave details of an explosion in the fuse-room at the Penhellick Fuse Works. Two

women were killed – Elizabeth Vivian (29) from Illogan and Ellen Opie from Pool.[21] The manager was named as Peter A. Renfree, and the number employed was given as thirteen women and girls, two men and a boy.

The explosion took place at the west end of the works. The fuse, winding, varnishing, tape and sump machine rooms were situated here, next door to a fitting-up shop. The manufactory was two storeys high. Five girls were in the winding room, which was separated from the fuse room by a wooden partition. The remaining six women were in the room below. Mr Phillips, the foreman and his assistant, helped by the boy, were in an outhouse placing fuses in tin cases, which were to be packed in casks for export.

At about 3.30pm a light bang caused by an explosion of gunpowder was heard, and shortly afterwards the building was ablaze. From the position in which the dead women were found, it was suggested that they had unsuccessfully tried to escape through one of the windows. Two other women were named as injured, namely Mary Blight of Tuckingmill and Elizabeth Jane Bayly of Camborne.

At the time of the explosion, there were 10 pounds of gunpowder in a small room adjoining the fuse room and about five pounds being used in the fuse room itself. The fire then took hold in the lower storey where 5,000 coils of finished fuse caught fire. The building was completely destroyed.

The *Royal Cornwall Gazette reported*: -

The bodies of Vivian and Opie were dug out of the ruins during Thursday evening, but they could not be identified except by measurement. About £150 worth of fuse was destroyed, and the total loss estimated at upwards of £300. During the past twelve months the company had contemplated erecting new works, having purchased property for that purpose about 400 yards from the building which has been destroyed. It is said that preparations have been made for the construction of the new manufactory. The old building was not insured, and the loss will therefore fall on the company.

The inquest was held at the *Plume of Feathers* in Pool, on Saturday April 29, with the Rev. W. Ellis as foreman of the jury. John Phillips, the engineer for the past eleven months, gave evidence. He said that the duty of the deceased 'was to attend upon the machinery, by which the safety fuse was made. The engine which drove the machinery was on the ground floor about forty feet away from the fuse room. Six machines were at work and they each had about a half-a-pound of gunpowder in its funnel'. Phillips stated that the stock of gunpowder was stored in a magazine about 400 yards away. He confirmed that the women had to wear nothing but leather shoes "without nails in the bottom".

The foreman of the jury asked some interesting questions: - 'There is an impression going abroad that you kept the girls locked up. Is that true?' Phillips replied 'No; there is not a lock in the place.' 'Where did the girls change their boots?' Phillips answered 'In the drying room upstairs.' He added 'the machinery was always carefully greased and oiled' so that he could not imagine the explosion occurred by friction. The Rev. Ellis then asked 'Did a train pass by at the time?' Phillips replied that he had not seen any.

Emily Hosking, a varnisher, stated that she had gone into the fuse room immediately before the fire, where the machines were not at work. Opie was dusting the machines with a small brush, while Vivian was brushing the floor with a long brush. The instant she left the room she saw fire. She said she had an old pair of slippers on her feet.

An Illogan carrier, by the name of Francis Cann, gave evidence. He saw a body on the ground floor when he looked through the north window. He poured water on the body and put out the fire. He then threw the trunk of the body out of the window into the garden of an adjacent dwelling-house; he found the head and body afterwards. Another body was lying about twelve feet away from the first one. The manager Peter Renfree thought that concussion must have killed the two women on the spot. He said he did not see any boots or shoes near the leg bones.

The foreman suggested that it would be much safer if the girls were to change their shoes downstairs in the future. Renfree said he would act on that suggestion. The verdict was 'Accidental Death'.

Ellen Opie (17) lived in Carn Brea Lane, and was the daughter of William Opie, a tin miner. They had previously lived in Tregajorran Lane, near the fuse works. Elizabeth Vivian was living in Illogan Downs in 1861, and was the daughter of copper mine labourer William Vivian. In 1861 she was a 'servant'. However, her sister Grace was a 'spinner' in a fuse factory; it seems likely she persuaded Elizabeth to join her.[22]

Peter Renfree was also an agent for the Scottish Life Equitable Life Assurance Society. Advertisements appeared regularly in the Cornish newspapers, giving his address as 'Penhellick Safety Fuse Works', although he lived in Redruth.

Figure 123. Enlargement of part of the 1924 aerial photo overleaf showing the fuse factory. © English Heritage

Figure 124. The Penhellick fuse works is to the left of, and below, the steam train. Taken in 1924. South Crofty in the foreground. © English Heritage

The Cornishman newspaper took great delight in reporting in July 1881 that Brunton and Co. had won a medal in the Melbourne Exhibition of 1881.[23]

THE GREAT SAFETY FUSE FIGHT

A keen contest is going on between the manufacturers of safety fuse in West Cornwall. No doubt great difficulty is experienced by the Companies in trying to do away with the monopoly which has been for so many years enjoyed at home and abroad.

Two firms of Cornish safety fuse manufacturers have won honours at the Melbourne Exhibition, official intimation having just been received that Messrs. Brunton and Co. of Pool, have been awarded a first class silver medal for their exhibits, while the older firm of Messrs. Bickford, Smith and Co. have secured a bronze medal.

Previous to this, however, Bickford, Smith and Company had carried off first prizes, and they always make quite a speciality of these samples for exhibit. No doubt the firm of W. Brunton and Co. are somewhat jubilant over their success at Melbourne, and will make the most of it in every way, and still try to beat down "monopoly".

What *The Cornishman* did not report, possibly because they had not known, was the fact that the prize-winning fuse had been made at the Brymbo works, Wrexham.[24]

The Penhellick Fuse Factory is closed

On the 31st of August 1898, Vivian Pearce, who was the Secretary to Bickford, Smith and Co. Ltd, wrote to Capt. Thomas, H.M. Inspector of Explosives, at the Home Office in Whitehall. His letter stated that Bickford, Smith and Co. Ltd. had acquired the fuse factory and premises at Penhellick. He enclosed the licences from Wm. Brunton and Co. and asked that they be transferred.[25]

The fuse works, together with all the machinery, had been assigned to Bickford's a few days earlier. The licences were endorsed by the Home Office by the 2nd of September 1898. Wm. Brunton and Co., Penhellick, was then closed down.

The buildings can be still seen in the 1924 photo, when several aerial photos were taken of industrial buildings in the vicinity. The fuse factory was almost destroyed in 1865, after the explosion. This was the replacement. All the buildings have since been demolished.

Peter Renfree died in June 1916, aged 80. He was living in Manor Villa, Green Lane, Redruth. William Brunton left India around 1858, which, according to his obituary, was prompted by an attack of rheumatism. He travelled to New Zealand, where he took a lease on a sheep-holding of 30,000 acres. In 1871, he became District Engineer of the railways in Southland, South Island, where he died on June

13th, 1881, aged 63. His obituary stated that he invented 'a fuse making machine of the most ingenious construction. This machine he did not patent, but kept it secret, and so effectively, that the process has never been divulged'.[26]

References, Chapter 11

1. Hughes, Steven. *The archaeology of an early railway system: the Brecon forest tramroads*. Royal Commission for Ancient and Historical monuments in Wales.
2. Morrison, T. A. Cornwall's Central Mines. The Northern District. p 204
3. *Ibid.*
4. Ancestry.co.uk website
5. *Glasgow Herald*, November 26, 1849
6. *Royal Cornwall Gazette*, July 12, 1850
7. *Royal Cornwall Gazette*, March 5, 1852
8. *Royal Cornwall Gazette*, March 5, 1852. p 8
9. *Royal Cornwall Gazette*, December 24, 1852
10. *Royal Cornwall Gazette*, July 1, 1853
11. *The West Briton*, June 5, 1902. Obituary of Oliver May, who died in Brymbo, North Wales.
12. *Royal Cornwall Gazette*, November 16, 1855
13. Dictionary of Australasian Biography (online)
14. *Royal Cornwall Gazette*, September 12, 1856
15. *Royal Cornwall Gazette*, January 18, 1856. p 4
16. *The London Gazette*, January 24, 1862
17. *Royal Cornwall Gazette*, May 8, 1857
18. *Royal Cornwall Gazette*, April 12, 1861
19. *Royal Cornwall Gazette*, April 19, 1861
20. *The Cornishman*, December 29, 1910
21. *Royal Cornwall Gazette*, May 5, 1865.
22. Census for 1861
23. *The Cornishman*, July 16, 1881
24. *The Wrexham Advertiser*, July 23, 1881
25. The National Archives UK. X71188
26. Institution of Civil Engineers, obituaries 1882.

CHAPTER 12

William Bennett, Roskear
1870-1924

An Engine Fitter

William Bennett was born in Gwinear in 1829. In June 1853 he married Catherine Lawry in St. Day. He was living in Redruth Highway and was described as an 'engine fitter'.[1] His father William was a farmer, while Catherine's father John was a miner. Eight years later the family was living in Davey's Row, Tuckingmill, when William described himself in the census as an 'engineer'.[2]

He had found employment with Bickford, Smith and Davey in their expanding fuse factory in Tuckingmill. In 1862, he had taken out a patent with William Bickford Smith. In the details of the patent he was described as 'Engineer' while his employer was described as 'Merchant'. The patent was for 'improvements in the method of, and apparatus for, preventing the injurious effects occasioned by smoke, sulphur and other gases' from chimneys and stacks'.[3] The Bickford-owned Camborne Gas Company was across the road from the fuse-works, and the bad smell from there might have caused a number of complaints. Both Thomas Davey junior and John Solomon Bickford were living close by, in Penlu Villas. The fuse works, too, had stacks – from the boiler, from the building where the gutta-percha was prepared, and from the pitch house. It seems that William Bennett was working also for North Roskear Mine, not far from his house. In September 1868 the engineer there 'W. Bennetts' was dismissed abruptly and replaced by Messrs. Michell and Jenkin of Redruth. The mine closed in 1874.[4]

He should not be confused with Captain William Bennetts, of Camborne, who was an agent at Wheal Grenville, Troon, at this time. The newspapers of the day invariably spelt the surnames of both men as 'Bennetts'.

The Start of the Roskear Fuse Factory

In 1869, William Bennett began building his fuse works on the old dumps of South Roskear Mine. The land was owned by the Pendarves Estate. Sales of copper ore were recorded from this old mine from 1819. In 1853 the sett was divided into two, when the eastern section became known as Pendarves Consols. This was worked

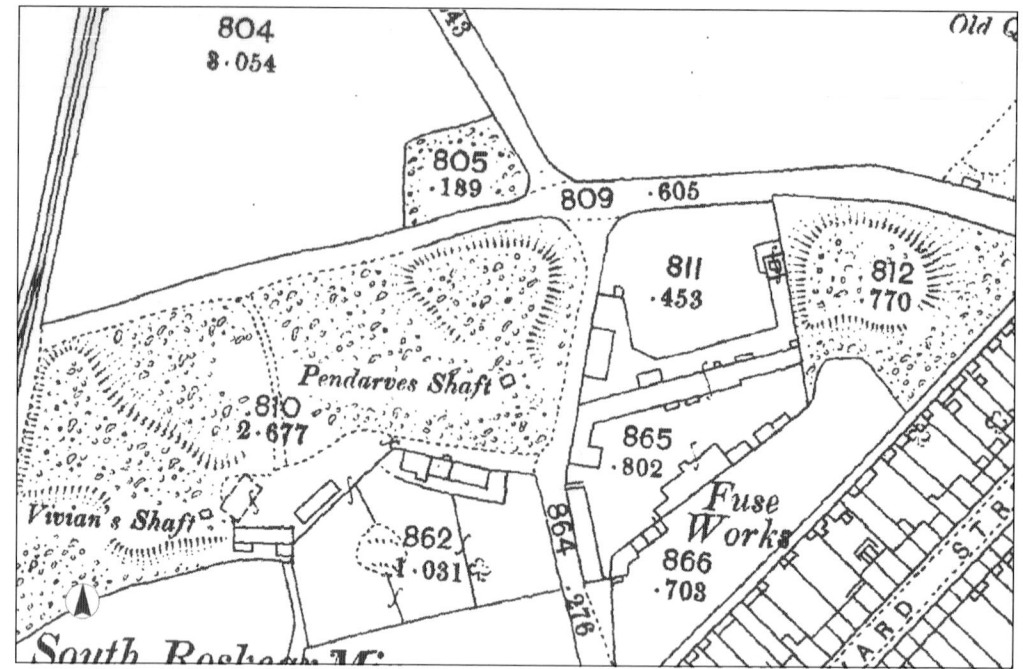

Figure 125. 1880 Ordnance Survey map of William Bennett's Fuse Works, Roskear

until late 1853, when the plant was offered for sale. It was re-opened in 1863 as 'Roskear Mine'. This cost-book company was wound up in May 1866.[5]

About three years after this, William Bennett leased the land on the eastern side of Roskear Row (now South Roskear Terrace) for his fuse works. An interesting description of the site at this time was given in a letter to the *Royal Cornwall Gazette* in August 1875 by an anonymous correspondent.

> I remember passing the site on which the works now stand about six years since, and it was then nothing but a dormant piece of mine waste, as a large quarry ran along one part, and in other parts thousands of tons of stuff stood in heaps. The stuff has been removed away at considerable expense, the quarry filled in, and the ground about is levelled.
>
> There is about three acres of ground enclosed by a permanent wall, and thereon are extensive and substantial buildings and offices erected for the manufacture of safety fuse. The works altogether show there was great taste displayed by Mr Bennett in laying them out. They must have undoubtedly have cost a large amount of money.[6]

The 1880 Ordnance Survey map shows a large amount of mine waste on the eastern side of the fuse works, and the whole factory seems to be enclosed.

By 1871, Bennett was describing himself as a 'civil engineer'. It was not until the 1881 census that he described himself as a 'Mining Engineer and Safety Fuse Manufacturer'. He is listed in the 1873 Kelly's Post Office Directory as 'Bennetts Wm, safety fuse manufacturer'.[7]

The order book or 'Journal' for the fuse works is in the Cornwall Record Office, and the first order is listed for April 1872. The tally of the stock of fuses began on April 15th.[8]

Coils	No.	Price in pence
3250	9	5
2100	1	3
500	7	5
500	12	8

The first order on the 25th of April was sent to J. K. Rogers, Liverpool, for 250 coils of No.1 fuse at 3 pence per coil and 750 coils of No. 9 fuse at 5 pence per coil. In October 1874 Bennett's works made a special order of No. 23 fuse. This was used, together with 2,000 pounds of Curtis and Harvey's 'extra strong' mining powder, to blast some part of a lime works at Minera, Wrexham, North Wales. In 1874, Bennett's were charging 1 shilling and 6 pence for a coil of No. 23 fuse. Their most popular fuse was their No. 9 fuse.

Expansion to the west side of Roskear Row
In 1872, South Roskear Mine re-opened, under the name of the South Roskear Tin and Copper Mining Company. Pendarves Shaft was cleared, and by April 1875 the mine was nearly drained to the bottom. In April 1879 operations at South Roskear Mine were suspended and the men started to construct dams in the levels to hold back the water in the western part of the mine. It was then planned to haul up the pitwork. Arrangements were made with the mineral lords to hold the sett until the price of tin rose to the point where they could resume operations. However, the machinery was put up for sale in the summer of 1881.[9]

At some stage after this, William Bennett expanded his works by going on to the site used by the Mining Company on the west side of Roskear Row. There may have been some old buildings there at the time. The yard on the western side of the site, beside the West Cornwall Railway siding to North Roskear Mine, was probably a coal yard, later replaced by a substantial two storey building.

In April 1888, Bennett applied for a patent for 'improvements in safety fuse'.[10] In May, two of his employees appeared before the East Penwith Petty Sessions in Camborne. The two young girls pleaded guilty to a breach of the Factory Acts, Rule 11, by leaving their workroom without changing their boots. The prosecution was recommended by the Government Inspector, Colonel Ford. He had suggested to

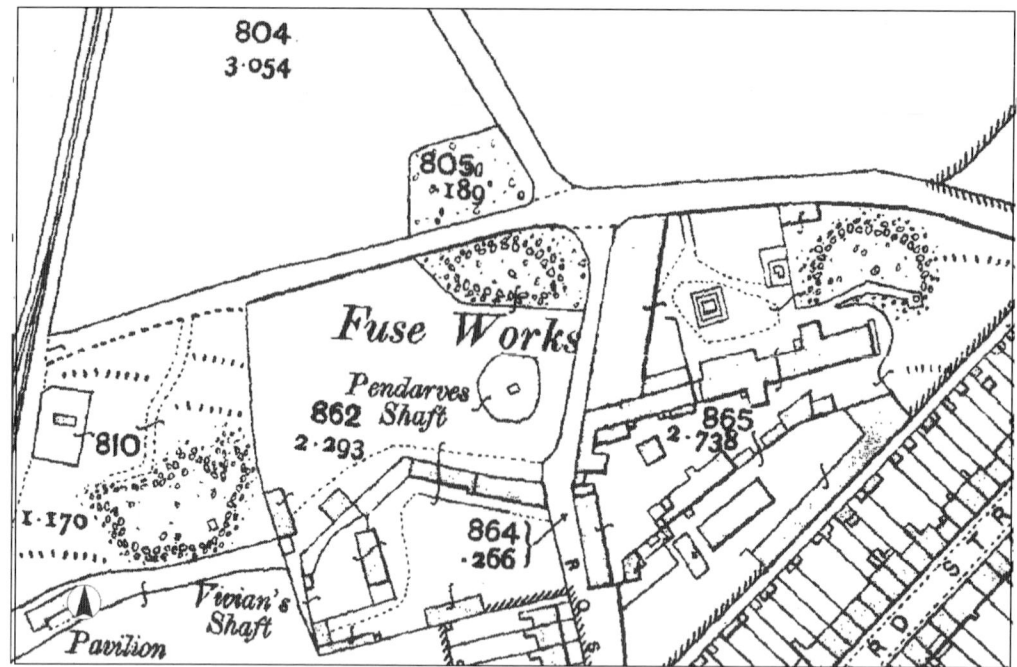

Figure 126. 1907 Map of William Bennett's Fuse Works, Roskear. Roskear Row runs N-S.

William Bennett that a 'strong caution' might in future prevent this, and the Bench agreed. The girls were given a few words of warning. They were named as Mary Williams and Annie Jane Williams. The latter was 17 years old, and lived in Tolcarne Street, Camborne.[11]

> WILLIAM BENNETTS
> ROSKEAR FUSE WORKS
> CAMBORNE, CORNWALL
>
> Safety fuses in all descriptions, for use in blasting operations at home and abroad; the best quality obtainable.
>
> Close prices and liberal terms to the trade. Buyers should inquire for BENNETTS' "CROWN BRAND" FUSE, which is always reliable. Samples and full particulars furnished on application.
>
> Figure 127 Advertisement in *The West Briton*. June 20, 1889

Around this time Bennett gave his fuses the brand name "Crown", and began advertising them under that name.

Breaches of Regulations

In December 1892, William F. Bennett faced four charges of breaches of Regulations at the Roskear Factory. Colonel Ford, H.M. Inspector of Explosives, had brought the

156

prosecution.[12]

- Allowing two barrels with 100 pounds of gunpowder in each to remain open in building K, instead of having enclosed hoppers of not more than 4 pounds each.
- Not taking proper precautions against fire by allowing a person to carry a barrel of exposed gunpowder from the adapting house P to fuse room K.
- Bennett had 203 pounds of gunpowder, instead of not more than 64 pounds, in building K.
- There were 450 pounds of gunpowder, instead of not more than 300 pounds, in another room.

Henry Grylls, prosecuting, said that it appeared to speak volumes for the way the factory was carried on that, when the Inspector came around after a long interval, he found that the gunpowder was handled so unguardedly as on the occasion of his (previous) visit.

The Bench of the East Penwith Petty Sessions dismissed the charge of having any knowledge or control of the men who carried the powder from one building to another, but convicted Bennett on the other three charges. He was fined, with costs. In the course of the trial, it transpired that the gunpowder was brought to the factory once a day by William Rowe, a carter, from the magazine at Trevarno, about one and a half miles away. This may have been an old magazine at West Wheal Seton.

FUSE

For Blasting with all kinds of Explosives.

WM. BENNETT SONS & CO.,

"CROWN BRAND."

Safety Fuse Manufacturers, CAMBORNE, CORNWALL

LONDON AGENT:
F. A. HILL, Baltic Chambers, Bishopsgate Street Within, E.C.

Figure 128. Bennett and Sons advertisement from 1887.

A Fire at the Roskear Fuse Factory

ALARMING FIRE AT CAMBORNE
SAFETY FUSE FACTORY ABLAZE

On July 5th, 1900, *The Royal Cornwall Gazette* carried this dramatic headline.[13] The blaze, which started after an explosion, destroyed the eastern section of the factory. A one storey building, 60 feet in length was ablaze, as well as some small buildings, which were not insured.

William Stephens, of the nearby Climax Rock Drill Works, ordered his employees to give what help they could, while several staff employed by Bickford, Smith and Co.'s fuse works arrived with their hose, buckets and wagons to help remove what they could. The vicar of Tuckingmill, the Rev. A. Adams, took a roll call of the women employed, which accounted for all of them.

The newspaper reported: - 'About four o'clock on Monday afternoon between 30 and 40 girls working in the four fuse-rooms were greatly alarmed by an explosion in their very midst. A quantity of gunpowder had ignited, probably by a reel or something from a machine dropping onto it, and in an instant the scene was one of indescribable confusion. Windows were blown out, and the place filled with smoke. The girls rushed into the yard, the various stores of powder for the supply of each machine exploding in all directions. Flames at once burst out through all the buildings and within a few minutes they were a mass of flames.'

The *West Briton* reported that the fuse factory was extending the works, and additional buildings were only partly erected, while others were very close to completion. They all suffered severe damage in the fire. It noted that about 200 people were employed.[14]

Report into the Fire

In September 1900, Major A. Cooper-Key, H.M. Inspector of Explosives, issued his report on the fire.[15] The blaze originated in the building G, half of which was being utilised for fuse spinning.

> This building measured internally about 145 feet by 18 feet, and was substantially constructed of stone with slate roof, and lined throughout with varnished match-boarding. The floor was of wood, with extra stout joists for the support of the machines. There were three regular exits, on each of three sides facing north, east and west, all opening outwards, in addition to which the windows, of which there was one behind each pair of machines, were swung on hinges at the top, so as to yield from very slight pressure from within.
> At a height of 10½ feet from the floor, a ceiling or half-deck of 1¼ inch

planks separated the gunpowder hoppers from the workpeople below. To these excellent arrangements may be attributed the almost miraculous c   of the work girls, of whom there were almost 40 in the vicinity.

Major Cooper-Key drew attention to the poor lighting in the half-deck room, stating that it would be impossible to see foreign matter in the hoppers. He also noted that the inspection of the work girls was not as thorough and efficient as it should have been.

> The clothing is not supplied by the firm, and whenever a girl comes to work in a new dress she is thoroughly searched by the forewoman to make certain that no pockets are worn, but there is no daily system of search.

The Inspector noted: - 'the use of wood as material for the lining of a building of this sort, where the risk is one of fire only, is strongly deprecated. It is practically impossible to prevent the accumulation behind the boards of fine powder dust, and the thorough and rapid manner in which the fire did its work may be attributed to this.' He advised a drencher, such as a perforated pipe, be installed and connected to an outside water tap.

Cooper-Key reported that Messrs. Bennett had always adopted all recommendations and stated: - 'it spoke volumes for their arrangements in the past that no case of serious injury to a working person by fire or explosion had occurred since the establishment of the factory (in 1871)'.

Death of William Bennett

In 1890/91 a severe influenza epidemic swept across Europe. Deaths were reported daily in the newspapers. William Bennett succumbed on May 2nd, 1891, aged 62.[16] His eldest son, Charles Frederick, was working as an engineer and mill manager in Dolgellau, Merionethshire, Wales, in an area that was producing gold. It was left to his sons William F. Bennett, then living in Roskear Villas, and Edward J. Bennett to take over the running of the fuse works. Edward was living in the Transvaal, South Africa at the time. His daughter Ida was born there in 1891, and another daughter, Catherine, was born there in 1893. His eldest son Thomas Edward had been born in Nicaragua in 1885. He returned to Cornwall, however, and in 1901 was living in Godolphin Road, Helston, where he described himself in the census as a safety fuse manufacturer.[17]

In 1903, William F. Bennett returned from British Columbia, on the West coast of Canada. He had obviously been successful in introducing the 'Crown Brand' gutta percha fuse to the Canadian market. A newspaper from British Columbia (quoted in *The Cornishman*) reported that the fuse had 75% of the entire market there.[18] The fuse works was enlarged at this time, by the erection of another building.[19]

William Bennett and Sons & Co. becomes a Limited Company

In July 1907 *The West Briton* accurately reported the formation of a new Limited liability Company, adding 'there will be no public issue (of shares), the re-arrangement being a family one'.[20] The Company (number 62766) was registered on June 26 1909, one month after Bickford, Smith and Co. Ltd was registered on the 14th of May.[21]

The nominal share capital was £36,000, divided into 1800 shares of £20 each. The 1800 shares were issued 'fully paid up otherwise than in cash'.

The Directors of 'William Bennett Sons and Company Limited' were registered as: -

William Francis Bennett. Roskear Fuse works, Camborne.
Edward John Bennett. Roskear Fuse works, Camborne

Shareholders in William Bennett Sons and Company Limited, 14th May 1909

Surname	Name	Address	Occupation	No. of Shares
Bennett	Charles Frederick	Victoria Lodge, Alton Road, Plymouth, Devon	Gentleman	1
Bennett	Edward John	Roskear, Camborne	Engineer	523
Bennett	William Francis	Roskear, Camborne	Fuse Manufacturer	523
Bennett	Thomas	Roskear, Camborne	Engineer	1
Pope	Annie Hatshill	Houser, Bickleigh Roborough, Devon	Wife of Joseph Pope, Engineer	1
Temby	Emily Jane	Roskear, Camborne	Wife of James Henry Temby, Engineer	1
Peter	William John	22 Basinghall Street, London EC	Chartered Accountant	745
Temby	James Henry	Roskear, Camborne	Engineer	5

On May 20th, 1908, 5 shares had been transferred to William John Peter by another shareholder to bring his total up to 745 (or 41% of the total). Annie Pope and Emily Jane Temby were the daughters of the firm's founder, William Bennett.

It appears that the shares held by W. J. Peter, a London accountant, were privately sold. At this time the two privately owned firms of Bickford, Smith and Co. Ltd and William Bennett Sons and Co. amalgamated.

It is worth examining the means whereby Bickford, Smith and Co. Ltd bought the Unity Fuse Works, Scorrier. In 1906 Sir George Smith and his son G. E. S. Smith purchased two-thirds of that Company, acting as Trustees for Bickford's. Five years later they purchased the rest, still acting as Trustees for Bickford's. In 1917 they assigned the Unity Fuse Works legally to Bickford, Smith and Co. Ltd. It may well have been the case that the same arrangement pertained here, and that selling the shares through W. J. Peter was done to keep the affairs of both companies private.

William Francis Bennett became the Managing Director of the Roskear Fuse Works, while Edward John Bennett was made a Director of Bickford, Smith and Co. Ltd.[22] The Roskear Company always kept the name William Bennett Sons and Co. Ltd. The buildings and land were not sold to Bickford, Smith and Co. until 1930. Amalgamation was complete by the time of World War 1. It is difficult to ascertain what each separate factory was producing at this time.

Figure 129. Nobel advertisement from May 1926.

Exemptions

At the start of December 1916, exemptions from enlisting into the armed services were sought by William Bennett Sons and Co. Ltd for five essential employees. One man was a fitter, two were coopers, another was a 'car man' and the last was the office manager. It was stated that five of the office staff had enlisted, and had been replaced by junior staff, who needed a high level of supervision. The fitter and the cooper (class A) were granted until January 1st. The other cooper and car man were granted until June 1st, and the managing clerk was granted until March 1st.[23]

War Work

The employees were engaged in filling 'gaines', or small booster fuses to insert into shells as an intermediary to cause the main explosive to explode. A building to contain the gaine-filling work was erected at the start of the war. Tetryl, used to fill the gaines, has several side effects for those that come into contact with it. (See Chapter 9). In his book *Cornish Explosives* Bryan Earl reported that many of the men working with tetryl at Roskear could not shave because of the pain caused by irritated skin.[24] The woman operatives meanwhile had formed a group called the

'Trelawney Pierrot Troupe'. At the start of 1917, the group gave a concert in the Godolphin Hall, Helston, in aid of the Red Cross.

In November 1919, buildings were being evaluated for rates by Messrs. A. Body and Sons, Plymouth. They discovered that the gaine-filling works had been partially erected within the limits of Bennett and Sons, and partially outside. They reported that the buildings had been demolished after the War.[25]

At the end of the war, Harry McGowan (who later received a knighthood) organised the merger of the British explosives companies. William Bennett Sons and Co. Ltd became part of the new company, to be called 'Explosives Trades Ltd.'. The Unity Fuse Company, Scorrier, which was then owned outright by Bickford's, was put into the new concern. It is highly likely that, by this time, Bickford, Smith and Co. Ltd owned all the shares in Bennett's. Fully paid shares were issued by Explosives Trades Ltd. for the shares of William Bennett Sons and Co. Ltd. The price at which the shares were exchanged is not known – Bennett's was under no obligation to put this information into the public domain.[26]

Figure 130 'The headgear, at the New Dolcoath mine at Roskear, where two were killed through a fall of roof'. *Western Morning News*, 7th August 1928. Bennett.

Communities were striving to return to normality after the war years. A concert was arranged in October 1919 in St. George's Hall, Camborne, with Camborne Town Band and the operatives of the Roskear Fuse Works. The Trelawney Pierrot Troupe again provided most of the entertainment.

Both Bennett's and Bickford's had to suspend operations in October 1919, because of the non-delivery of materials caused by a strike by railway workers. This was resolved, and both works resumed operations in the middle of October.[27]

The following year, in October 1920, *The West Briton* reported that the coal miners' strike was having a very detrimental effect on the Tuckingmill and Roskear Fuse Works. Bennett's had enough coal stocks by the end of October to last only another week or ten days. The same newspaper also carried very depressing news about the local mines, which were ceasing work, never to re-open.[28]

Sir George J. Smith died in October 1921. He was the managing Director of Bickford, Smith and Company, and Vice-Chairman of Nobel Industries Ltd. (the former Explosive Trades Ltd.). His son G. E. Stanley Smith CBE took over as Managing Director of the Bickford, Smith and Co. Ltd. fuse works.[29]

Figure 131. The New Dolcoath Mine and old fuse works buildings, 1924.

'GOOD NEWS FOR CAMBORNE'

The Cornishman printed this optimistic headline in September 1922.[30] The Directors of Dolcoath Ltd announced plans to re-open the old South Roskear Mine. The mine was to be called 'New Dolcoath'. Bennett's Fuse Company had erected buildings on the site about twenty years previously, and these had to be purchased from Nobel Industries Ltd. In May 1923 the price was agreed at £1,350 and 500 shares in the new venture. In July the bungalow that was on the site of the old South Roskear Mine, together with the adjoining yard, were included in the sale. The leaseholder was William Cole Pendarves, owner of the Pendarves Estate, who also owned the mineral rights.

Such was the excitement engendered by this revival in Cornish mining that the local newspapers covered the progress of the mine in great detail. Camborne photographer William John Bennett took photographs of the mines progress, especially the erection of the 80 feet high steel lattice headgear (from William's Shaft, Dolcoath) on the New Roskear Shaft from 1924. These photographs fortunately survive, and in the background can be seen the old fuse works buildings. The photograph in Figure 130 appeared in a newspaper after two men had been killed by a collapse of ground at the 2,000 feet level.

Several aerial photographs dating from 1924 exist. The photos were taken principally of the New Dolcoath Mine. In the photos can be seen a large two storey building (at roughly a right angle to what is now South Roskear Terrace) which was, at one time, used by the fuse works. Stairs to facilitate rapid escape can be seen at the rear. The long low building beside the new square stack (and backing onto Camborne Cricket Ground) may have been used as offices by the fuse works. One wall consists of numerous windows, which does, however, suggest the building was used in the fuse making process. The two storey white building parallel to, and backing onto, the train tracks may have been used as a storage depot for goods out and in. The fuse company was receiving large shipments of jute from Dundee, by ship to Hayle, and they may have been unloaded here. The date of this substantial two-storey building may be c.1904, when the fuse works was expanding. No trace remains of the building today. In the aerial photo, the building with the ramp is the new boiler house.

On the east side of Roskear Row, the 1924 aerial photo shows the fuse buildings in good repair, although devoid of any sign of activity. Work was abandoned on New Dolcoath in 1929, and in 1936 the shaft was taken over by South Crofty Mine. The site is now abandoned, completely overgrown with vegetation and derelict.

Figure 132. Pendarves House. The roof was removed in the mid-1950s.

Death of W. F. Bennett

William Francis Bennett, the son of the founder of the Roskear Fuse Works, died towards the end of February 1929, aged 70. The *Western Morning News* noted: - 'until the amalgamation with Messrs. Bickford, Smith and Co. Ltd, the late Mr. Bennett was a Director of the firm of Messrs. William Bennett and Sons Ltd., safety fuse manufacturers, Roskear'.[30]

Sale of the Pendarves Estate, 1930

The Pendarves mansion was a few miles south of Camborne, between Killivose and Treslothan. Like all large estates that drew their principal revenues from the declining mining activities on their land, it was finding it difficult to survive.

William Cole Pendarves died on Monday May 13th 1929, aged 87. The estate passed to his son, John Stackhouse Pendarves.[31] The lands, farms, buildings, ground rents and reversions were put up for sale in May 1930. Messrs. E. Mitchell and Sons, Auctioneers, Penzance, held the sale over four days – June 16-19th, in the Commercial Hotel, Camborne.[32]

A 'reversion' is a legal term. There was quite often a 'reversion' clause in leases of that period. This meant that, if the lease was not renewed by the lessee, for whatever reason, the leasehold land and the buildings thereon reverted to the lessor – in this case the Pendarves Estate.

> **Preliminary Notice**
> PENDARVES ESTATE
> In the Parishes of Camborne, Crowan,
> Illogan, Redruth, Gwennap and St. Erme,
> CORNWALL
> FREEHOLD PROPERTIES
> Comprising
> FARMS, SMALL HOLDINGS, COTTAGE
> HOLDINGS, COTTAGES, ACCOMMODATION
> LANDS, GROUND RENTS and REVERSIONS,
> Embracing a total area of about
> 1,630 ACRES
> TO be offered for SALE by AUCTION
> shortly (unless previously disposed of
> by private treaty), on a date to be
> announced, in 818 Lots.
> Land Agents and Surveyors
> Messrs. GLANVILLE, HAMILTON, and
> WARD,
> 53, Lemon-street, Truro
> Solicitors,
> Messrs. BURGESS, TAYLOR, and
> TRYON
> 1, New-square, Lincoln's Inn,
> London, W.C.2.
>
> **Figure 133.** Sale notice in *The Western Morning News*, April 9th, 1930

It is worth noting that the sale of Roskear Fuse Works took place 60 years after William Bennett first leased the site from the Pendarves Estate in 1870. Sixty years was quite often the term of a lease. William F. Bennett, the son of the original lessee, had died in 1929. It seems that the lease was not renewed, and the site and buildings reverted to the lessor.

On Thursday June 26th 1930, *The Cornishman* carried the following report: -

The sale was continued on Wednesday, when Messrs. Bickford, Smith and Co. Ltd of Tuckingmill purchased the fuse works at North Roskear for £500, and an explosive store and 2 acres of land at the same place for £25.

There is no evidence that Bickford, Smith and Co. Ltd (which was then owned by ICI) ever used the works again for the manufacture of fuses. The buildings have been the target of an arson attack and the site is now derelict.

References, Chapter 12

1. Cornwall Online Parish Clerk
2. Census 1861
3. *The London Gazette*, May 9, 1862
4. Morrison, T. A. *Cornwall's Central Mines. The Northern District.* Chap. 20
5. Ibid.
6. *Royal Cornwall Gazette*, August 29, 1875
7. Census for 1871 and 1881
8. Journal, Bennett's Fuse Works. CRO. X1302/1
9. Morrison, T. A. op. cit.
10. *Bristol Mercury*, May 2, 1888
11. *The Cornishman*, May 20, 1888
12. *The Cornishman*, December 29, 1892
13. *Royal Cornwall Gazette*, p 4. July 5, 1900.
14. *The West Briton*, July 5, 1900
15. *Royal Cornwall Gazette*, September 27, 1900
16. *The Standard*, May 23, 1891
17. Census 1891
18. *The Cornishman*, August 13, 1903
19. *The Cornishman*, p 4. September 9, 1903
20. *The West Briton*, July 25, 1907
21. Cornwall Record Office. BY/339
22. *The Cornishman*, October 19, 1933

23. *The West Briton*, December 4, 1916
24. Earl, Bryan. *Cornish Explosives*. 2nd edition. p 137. The Trevithick Society. 2006.
25. Cornwall Record Office. BY/323
26. *The Cornishman*, October 29, 1919
27. *The Cornishman*, October 15, 1919
28. *The West Briton*, October 28, 1920
29. Advertisement from *Western Morning News*, May 3, 1926
30. *The Cornishman*, September 20, 1922
31. Buckley, Allen. Dolcoath Mine. A History. The Trevithick Society. 2010. p391
32. *Western Morning News*, February 23, 1929

Figure 134. The two storey building (centre) and the buildings at rear were part of Bennett's fuse works. Royal Institution of Cornwall

Figure 135. This building backs onto the railway and may have been used as a store by Bennett's. RIC

Figure 136. This long single storey building backs onto the cricket ground. (Note Roskear Church) Royal Institution of Cornwall

Figure 137. The building used by Bennett's fuse works. Notice all the windows. Use unknown. RIC

Figure 138. William Bennett's Fuse Works, Roskear. Looking west. 1924.
© English Heritage

Figure 139. William Bennett's Fuse Works, Roskear. Looking east. 1924.
© English Heritage

PART 3

THE REDRUTH AREA FUSE WORKS

923 Springboks press—1st half.   Springboks v. Cornwall at Redruth, 1912.

CHAPTER 13

The British and Foreign Safety Fuse Company, Redruth
1846-1913

William Bickford's patent for the safety fuse was granted on the 6th of September 1831, and was valid for 14 years. Immediately on expiration of the patent a fuse factory was established in Redruth, at the northern side of the Redruth Brewery (1742). The Brewery Leat, an important source of water power, probably influenced the location of the factory.

The manager was William Henry Launder (26), who was listed in the 1841 census as a miner from Tuckingmill. The funds were provided by John Charles Lanyon (45), who had an established business as a currier in Fore Street, Redruth. Lanyon may have had other backers in the project, namely John Hocking, an engineer, Michael Morcom, a St. Agnes mine agent, and Thomas Garland.[1] However, it appears that it was William Launder's wife, Sarah Rowe, who had all the expertise. At the time of their marriage in Camborne in February 1841, she is listed as a labourer in a safety fuse factory.

In 1851, William Launder (30) was living in Roach's Row, Redruth and was described, as was his wife Sarah, as "Safety Fuse Manufacturer". Their first advertisement appeared in the *Royal Cornwall Gazette* on April 24th 1846. The short space of time between the expiry of Bickford's original patent and the appearance of the advertisement would indicate that the fuse factory was making basic safety fuse using easily manufactured machinery, and was probably supplying local mines. North of the site was an upland area known as 'Tolgus Downs', where several copper mines were situated.

> Safety Fuze for Blasting Rock,
> Submarine Explosions &c.
>
> The British and Foreign Safety Fuze Company beg to inform the Managers and Agents of Mines, and other parties engaged in works requiring the SAFETY FUZE, that they are now able to supply that article in any quantities, and of such descriptions as be required.

> The B. and F. Safety Fuse Company have spared no expense in order to make an article of the first quality, and they hope, by strict attention to business, to merit a continuance of the orders they may be favoured with.
>
> Orders from any part of the kingdom will be executed with every possible dispatch; and particular care will be observed in packing Fuze, which may be wanted for exportation.
>
> Dated Redruth, 21st April, 1846
>
> **Figure 140.** Advertisement in the *Royal Cornwall Gazette*, April 24 1846

Advertisements appeared in Glasgow (1847), Dublin (1847), Wexford (1848) and Newcastle (1848) where it was advertised as "The Cheap Safety Fuze".

By 1851, John Charles Lanyon was still living in Fore Street, Redruth, but was now describing himself, in the census, as 'Merchant, ship owner, tanner and fuse manufacturer'. He had an ironmongers' shop there. The fuse factory appeared to be a small part of his business. He would have been supplying the many Redruth area mines with goods from his shop, and also supplied them with fuse. In the fuse factory he employed about six young female workers plus William Launder and his wife Sarah. The youngest operator was Ann Jones, aged 12, from Tolgus Downs. The clerk was Pearce Vaughan, aged 40, who resided in Ford's Row. The Launder family had moved into a house on nearby 'Roach Row', where they lived with their three young children and a servant.

Slater's Directory of Berkshire, Cornwall and Devon for 1852-53 listed three Redruth Safety fuse manufacturers.

BRITISH AND FOREIGN SAFETY FUSE CO. Redruth (John C. Lanyon, Manager)
BRUNTON, William & Co., Penhellick, near Redruth
HAWKE, Edward & Co., Scorrier, Truro

The 1861 census for the area shows 22 female safety fuse operatives, plus William Launder and his sister-in-law. The youngest were Mary J. Nettell, aged 11, from Plain-an-Gwarry and Mary Anne Eade, aged 12, from Wentworth. In September 1861, a fire broke out in the varnishing shed, due to varnish boiling over in the absence of the workpeople at dinner time. The whole of the wooden roof was destroyed, but the machinery suffered little damage. The shed was some distance from the fuse works, and fortunately the fire did not spread.[2]

John Charles Lanyon died on the 24th November 1868 aged 67. His ironmonger's shop in Fore Street was taken over by his son Alfred, then 22 years of age. The management of the fuse factory was in the hands of William Launder, who by this time was living in Rose Hill, close to the factory. He had his sister-in-law Mary

Rowe (37) living with him and his family. They both describe themselves in the 1861 census as 'Safety Fuse Manufacturers, Employers'. His wife Sarah (41) was mother to eight children. Rose Hill is on the western edge of a very old established area, known as Plain-an-Gwarry. By 1839, the main thoroughfare, called Plain-an-Gwarry, was almost completed, and between 1830 and 1850 the terraces of workers cottages were constructed.[3] Today this area lies within the Cornish Mining World Heritage Site. The Plain-an-Gwarry was a banked amphitheatre where plays, particularly 'miracle plays' were performed in Cornish, and it is thought to be at the eastern end of the area, where Green Lane and Drump Road meet. In the 1880's Edward Tangye started a small fuse works here, to the rear of his house known as 'The Elms'.

In October 1865, the British and Foreign Safety Fuse Company won a medal for their 'miners' patent safety fuse'. This had been entered into the International Exhibition held in Dublin. Bickford, Smith and Co. also won a medal for their 'patent safety fuse'. These awards were reported widely in the Cornish newspapers and appeared on the price lists of the firms.

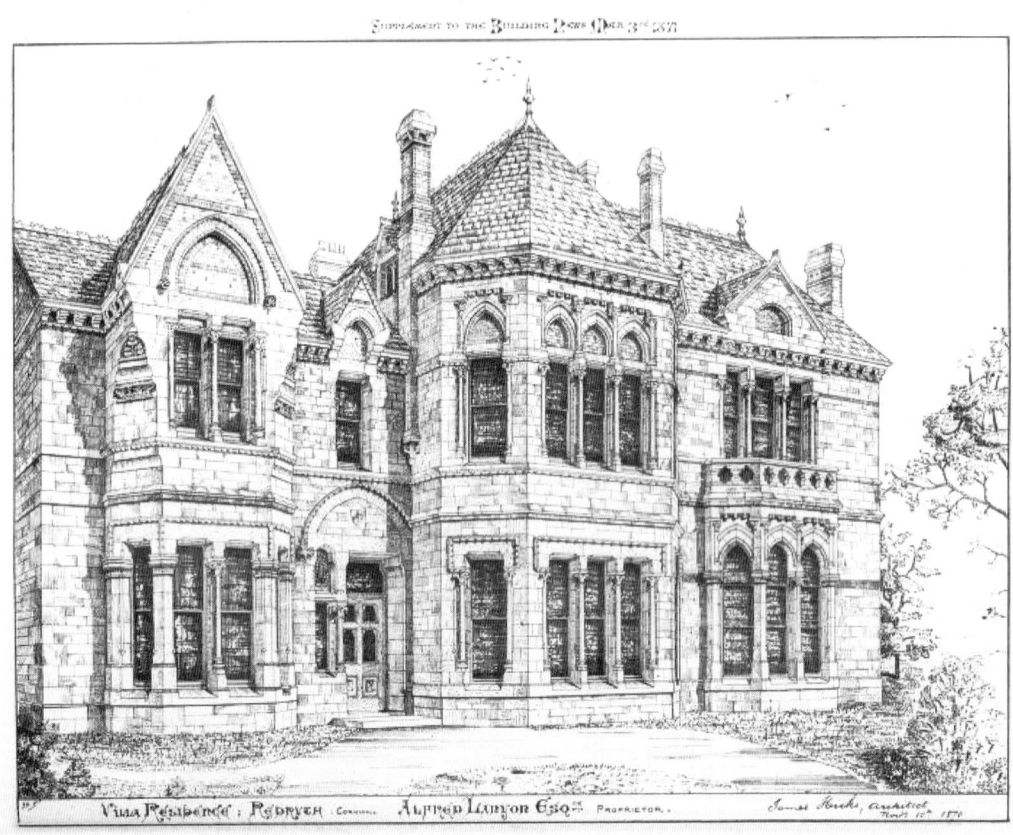

Figure 141. Tolvean House, Redruth, the home of Alfred Lanyon
British Builder, March 3, 1871

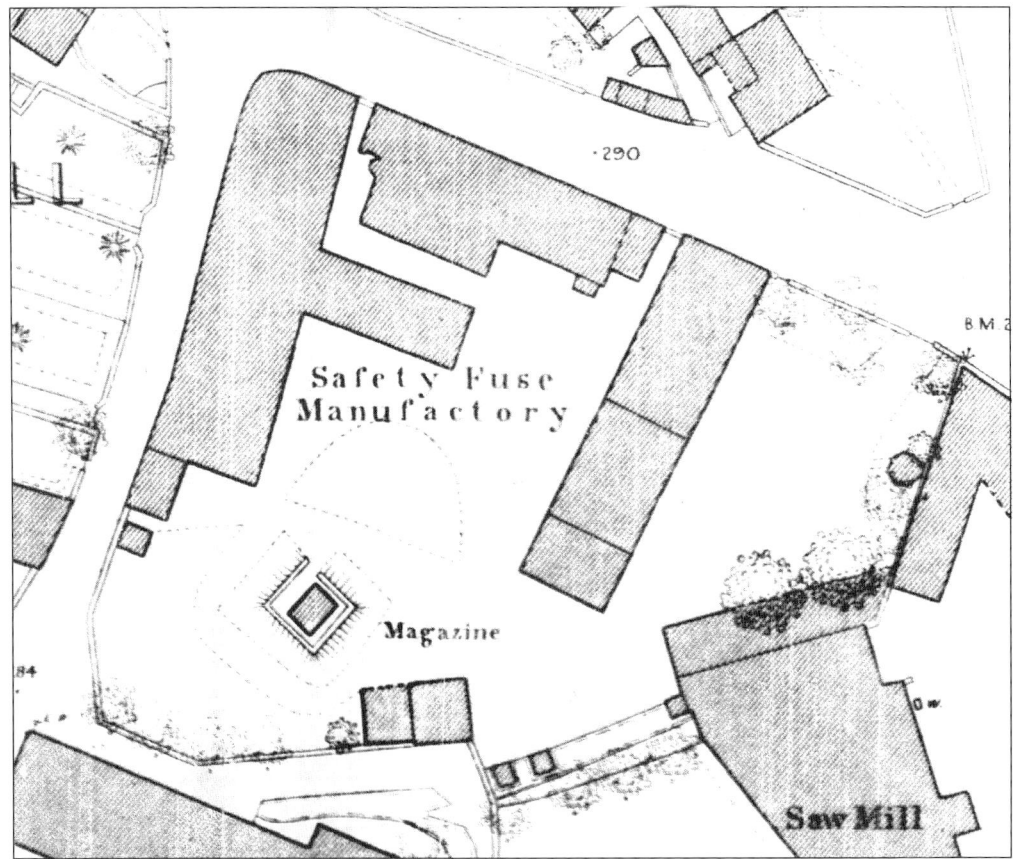

Figure 142. 1880 Ordnance Survey map of the British and Foreign Fuse Works.

Doidge's Directory for 1866, for Redruth, lists William Henry Launder as 'resident foreman' in the fuse factory. He and his family had moved into the fuse works, where he is listed in the 1871 census as resident in the 'Safety Fuse Yard'. His son, John Launder, died in June 1874, aged 26. In the same Directory, Alfred Gee, who lived on the west side of Green Lane, was listed as a Commercial Traveller for the 'Redruth Safety Fuse Co.'

In Kelly's Directory of 1873, Alfred Lanyon was living in Tolvean, a large mansion he had built in the West End of Redruth. The architect was James Hicks. He was listed as Manager of the fuse works, president of the Gas Works and Vice-Chairman of the School Board, as well as running his ironmongers business in Fore Street. William Launder and his family were noted as living in 'Fuse Cottage'.

In March 1875, an application was made in the London Court of Bankruptcy for an injunction to restrain certain proceedings in bankruptcy in Sheffield, which had been instituted by Messrs Lanyon, Hocking and Francis, trading under the name of the British and Foreign Safety Fuse Company as fuse manufacturers and merchants. The gentleman in financial trouble was Mence Wilkinson, formerly a chemist in

Figure 143. Advertisement for The British and Foreign Safety Fuse Co. from *Kelly's Directory* 1889

Sheffield and a farmer at Hurlfield, Yorkshire, but who lived in London at that time. Wilkinson was declared bankrupt, however, by the Sheffield Bankruptcy Court. What goods he had received from the fuse factory was not stated.[4]

In May 1875, in order to comply with the Explosives Act, 1875, the British and Foreign Safety Fuse Company applied for assent from the Justices for the East Division of Penwith for a magazine for explosives on a site in Tolgus. They also applied for a licence, as required under the Act.[5] The magazine may have been at Great South Tolgus copper mine. When the mine was abandoned in 1871, all the mine plant was purchased by J. C. Lanyon and Sons for £2,308.[6] Unusually, in the 1911 census for the area, a 'powder magazine' was listed on Tolgus Downs, which would seem to indicate that it was still in use.

Sometime after this William H. Launder left the fuse works in Redruth, and travelled to Swansea, where he was involved in starting the Swansea Safety Fuse Company at Pipe House Wharf. In the 1881 census, he and his wife Sarah were living in Walter's Road, Swansea; he gave his occupation as 'Safety Fuse Manufacturer' employing 36 girls and 2 men. His son William (36) was listed as the Manager of the fuse works, while another son, Richard, 25, was an agent for the fuse company. The company was taken over in 1897 by Nobel's Explosives Company Limited, at which

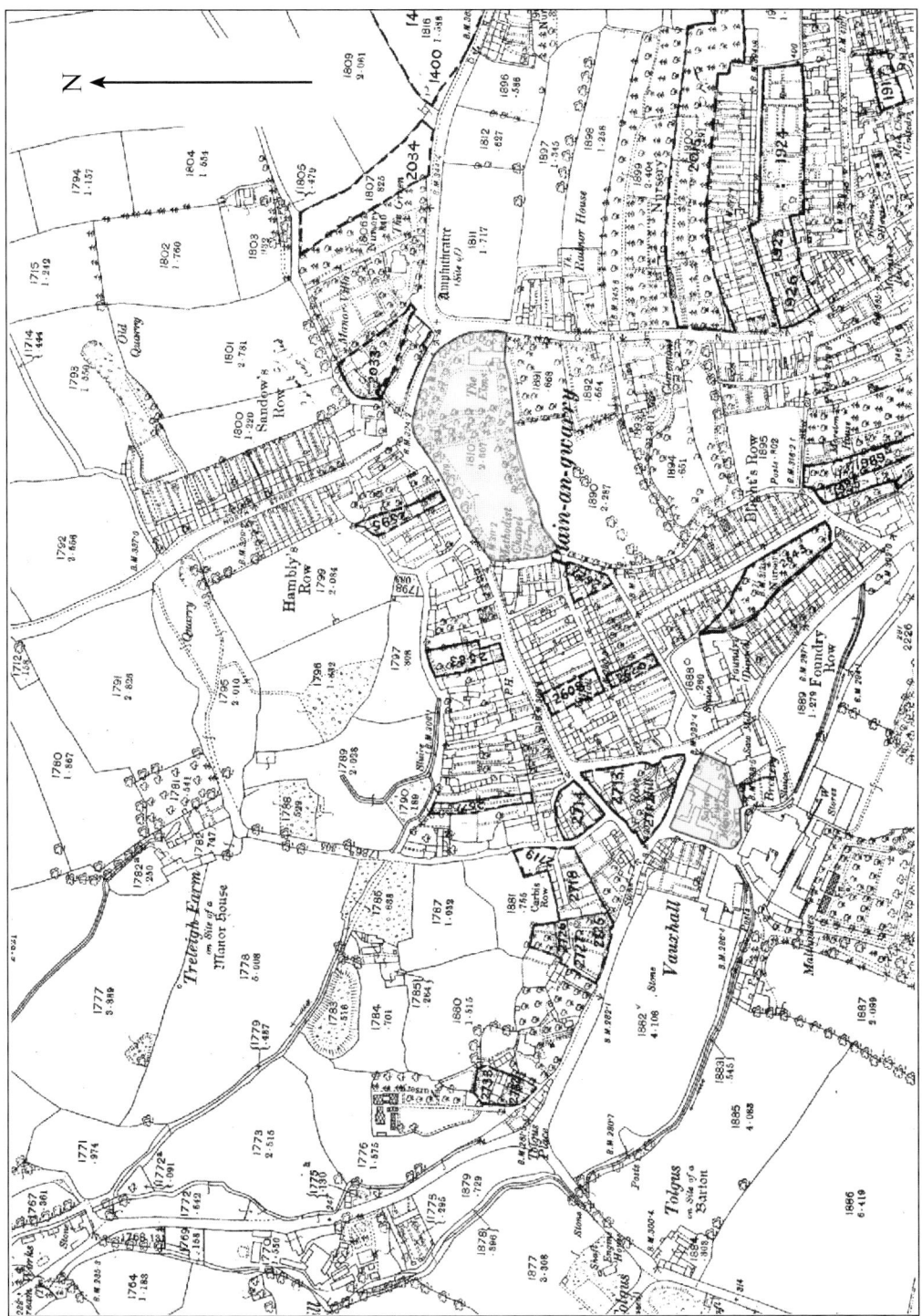

Figure 144. North Redruth, 1880. British and Foreign fuse works and Brewery (right of centre) and The Elms (Tangye's) (above centre).

time Launder was the managing director.

In 1878, Alfred Lanyon was granted a patent (No. 4792) 'for improvement in the safety fuse'. The British and Foreign Fuse Company won a bronze medal for their fuse in December 1882. They exhibited at the Annual Exhibition of the Cornwall Mining Institute, held in the Assembly Rooms, Camborne.[7] Earlier, in September, their samples were 'commended' at the Royal Polytechnic Society Exhibition in Falmouth.[8]

In July 1883, Alfred Lanyon took his fuse workers on an outing to Falmouth and Truro. The *Cornishman* noted the trip was at 'the Company's expense'.[9] In May 1884, the Company produced their new trade mark. The designer is not known. They advertised a tri-colour thread of red, white and blue running through their fuses.[10]

In October 1884 the Redruth Liberal Association was inaugurated, with Alfred Lanyon as the President.[11] He again treated his employees to an outing in 1886, this time to Perranporth. The foreman in charge of the group was named as James Thomas.[12] In the 1881 census he was described as an engineer, with an address in Green Lane. In the same census there were approximately 22 females working in the factory, as well as a cooper named William Heyden (54) and his son John (22), a carrier for the company. The accountant was William Manley, 23, who lived on South Row.

In July 1887, a competitor started a fuse works in Plain-an-Gwarry. His name was Edward Tangye, and he was a former member of the famous engineering firm of Tangye Bros, Cornwall Works, Birmingham.

Alfred Lanyon and some colleagues decided to diversify. In December 1889, notices began to appear in various newspapers advertising shares for sale in a new Company called 'The British and Colonial Explosives Company, Limited'. The capital was £100,000 in 100,000 shares of £1 each. The five Directors were named as:-

Colonel J. Lowther du Plat Taylor. CB

Edward Boyle Esq. 1 King's Bench Walk, Temple, EC.

Alfred Lanyon Esq. Merchant & mine-owner, Redruth

Thomas Pryor Esq. Mine-owner, Redruth

Captain Josiah Thomas, M.E., Dolcoath, Camborne.

The company was formed to manufacture all kinds of explosives, but principally dynamite. The prospectus stated that the company had acquired a freehold site of 100 acres on the north coast of Cornwall.[13] The site was north-east of St. Agnes, near the cliffs west of Perranporth. It was acquired by Nobel's Explosives Company in 1893.[14]

Edward Tangye's fuse factory was established by the time of the 1891 census, at The Elms on the eastern edge of Plain-an-Gwarry. Thirty people in the 1891 census

Figure 145. Aerial photo (enlarged) from 1924 showing the Redruth Brewery, with the old British and Foreign fuse works to the rear. ©English Heritage

worked in these two local fuse factories. William Heyden and his son John were still employed by the British and Foreign Safety fuse works, where the engineer was James Thomas. However, by the next census, (1901) all the Redruth mines had ceased to operate, leading to large-scale emigration and a general depression in the area. In the 1901 census, only the British and Foreign safety fuse factory was operating. In the census, there were only 11 female employees listed. William Heyden (73) was listed as a 'packer, safety fuse' while his son John (42) was listed as 'foreman'. Dolcoath Mine in Camborne was still producing tin, however, and in 1895 Alfred Lanyon had 31 shares and Edward Tangye owned 47 shares, with his brother Joseph holding 31.

Alfred Lanyon was still taking his employees on an annual outing, and *The West Briton* gave a very full account of their August day out in 1900.[15] The number of employees seems very large – but many of the women would have brought family members with them.

> The annual excursion of the employees of the safety fuse company at Redruth took place on Wednesday at Portscatho. Between forty and fifty of the employees left Redruth at 7 a.m. in charge of Mr. J. H. Heyden, and were conveyed to Portscatho in a Jersey car and brakes, each drawn by four horses, the arrival being about 11a.m. As the weather was fine throughout the day, the drive was greatly enjoyed, especially the crossing of the Fal at King Harry Passage by the steam ferry. Directly after the seaside was reached, lunch was partaken at the Plume of Feathers. Dinner followed about two p.m., and tea at six o'clock, everything being provided by the proprietor of the works – Mr. A Lanyon, J.P, C.C. The day's outing was filled up by the usual seaside

recreations, including boating, bathing and rambles along the cliffs. Home was reached about 11 p.m., all agreeing that the day had been most happily spent.

In 1904 Bickford, Smith and Co., Ltd, together with Nobel's Explosives Company Ltd bought the British and Foreign Fuse factory in Redruth. Nobel's purchased one third, while Bickford's bought two thirds. Two years later they bought Nobel's share.[16]

Bickford's kept the fuse factory in Redruth going for a few years. In 1908, the fuse works was given a rateable valuation by Redruth Union.[17]

REDRUTH (1908)	OLD	NEW
Redruth Brewery	£331	£1,064
-bonded stores	45	100
-stables, stores &c	-	50
-waterworks & pipes from Seleggan to Brewery	47	95
W. E. Bacon Co.'s factory	84	420
Cornish Tin Smelting Co.'s works	135	350
Gas Co.'s works, mains &c	200	343
Tabb's Hotel	191	335
Smith's boot factory	56	260
Foundry Co.'s foundry	52	210
Redruth Tin Smelting Co.'s works	68	200
British & Foreign Fuse Co.'s factory	55	164
W. William's bacon factory	34	116
Buller's Arms	68	112
London Hotel	76	108
King's Arms	79	100

It seems that Bickford, Smith and Co. closed the fuse works a few years after this, probably in 1913. In the 1911 census, taken on April 2nd, John Heyden is listed as residing in a 'Private House' in the Fuse Factory. He described himself as 'Fuse Works Manager'. The factory does not appear on the 1908 Ordnance Survey map of Redruth, which reflects its loss of importance. In February 1912, the death was reported of a woman who had worked at the British and Foreign Fuse Company for 44 years, namely Miss Catherine Sparnon (54). The *West Briton* reported that twelve female employees acted as bearers.[18] Whether these represented the total number of operatives was not stated. By contrast, in 1909, Bickford's employed 280 women and 12-15 men. In March 1913 Alfred Lanyon disposed of the Cornish Tin Smelting Company, which he sold to a London Company.[19]

Figure 146. A man working on the stack of the British and Foreign safety fuse works at Plain-an-Gwarry. Paddy Bradley

Figure 147. The two young girls are standing on the corner of Rose Hill and Roach's Row, across the road from the fuse works. C 1910. Bragg

Figure 148. View of Redruth, looking north. 1924. Fuse works & Brewery are top left. Tangye's (The Elms) is top right. © English Heritage

Figure 149. Former fuse works building at Vauxhall, converted to offices for the Redruth Brewery. Since the closure of the brewery this has been gutted by fire. Peter Joseph.

Figure 150. Fuse works stack at Vauxhall. This was demolished in c.2008 for alleged safety reasons. Peter Joseph.

In August 1913,[20] a Monsieur Kinsman, a French fuse manufacturer, visited Bickford, Smith and Co. Ltd. in Tuckingmill and Redruth. He noted; - 'the Redruth works is a smaller one, acquired by purchase from Mr. Alfred Lanyon'. Bickford's were enlarging their factory in Tuckingmill at this time, including building a jute spinning building. They probably had no need for another small fuse factory, and closed down their Redruth works. Alfred Lanyon died on the 5th of March 1915, aged 80. [20]

References, Chapter 13

1. CRO document STA/693/1046. 1857. Plaintiffs in a case were Lanyon, Hocking, Garland and Michael Morcom (a St. Agnes mine agent) who were described as 'safety fuse manufacturers of Redruth.
2. *Royal Cornwall Gazette*, September 27, 1861
3. Plain-an-Gwarry Conservation area. CCC March 2010
4. *Sheffield Daily Telegraph*, March 4, 1875
5. *Royal Cornwall Gazette*, June 3, 1876
6. Morrison, T. A. *Cornwall's Central Mines. The Northern District.* p 80.
7. *Royal Cornwall Gazette*, December 15, 1882
8. *The Cornishman*, September 7, 1882
9. *The Cornishman*, July 26, 1883
10. *Western Times*, May 5, 1884
11. *Royal Cornwall Gazette*, October 17, 1884
12. *Royal Cornwall Gazette*, September 3, 1886
13. *The Cornishman*, December 5, 1889
14. Earl, Bryan. *Cornish Explosives*. The Trevithick Society. 1976
15. *The West Briton*, August 16, 1900
16. See Reference 1, Chapter 3. G. L. Wilson
17. *The West Briton*, October 22, 1908
18. *The West Briton*, February 15, 1912
19. *The Cornishman*, March 13, 1913
20. *The Cornishman*, August 21, 1913

CHAPTER 14

The Unity Fuse Works, Scorrier
1846-1918

On the 5th of August 1828, Edward Henry Hawke married Charlotte Pinwell in Thaxted, Essex. She was the daughter of John Pinwell, a Purser in the Royal Navy. They returned to Hawke's birthplace of Gwennap, to live in Tolgullow, a hamlet north east of St. Day and about two miles east of Redruth.[1]

Their first daughter, Charlotte Mason, was baptised on May 31, 1830, when Edward Henry described himself as a 'roper'. When their next daughter, Julia Pinwell, was baptised on February 9 1831, Edward described himself as a 'ropemaker'. Edward senior described himself as a 'rope manufacturer' when his only son Edward Henry was christened in 1832.[2]

It was at his rope manufacturing place that Edward Hawke started a small fuse works. The machinery to spin a rope and to spin a fuse would have been somewhat similar, and the large number of copper and tin mines surrounding his premises would have guaranteed regular customers.

Writing in 1909, an anonymous contributor to *The Rise and Progress of British Explosives* stated that from the commencement "the works have been under the control of shareholders and management in the Cornish mines, with mining engineers".

William Bickford's original safety fuse patent expired towards the end of 1845, and this allowed competitors such as Edward H. Hawke to enter a new and profitable industry.

To ENGINEERS, ARCHITECTS
CONTRACTORS
GREAVES LIAS LIME AND CEMENT

LIAS LIME, as Mortar, is especially adapted for sewers,

dock walls, hydraulic, railway, and other works; its adhesive qualities being superior to Roman cement.

As CONCRETE, it has more power than another cement or lime, as it forms a solid mass - sets hard as rock under water – using one bushel of lime to one of

> gravel.
>
> LIAS CEMENT is easily worked, of a beautiful stone colour (similar to Portland stone), does not vegetate or crack, hardens by exposure to the atmosphere, and is well adapted for modelling any casting.
>
> J. THOMPSON, agent office, St. James's Chambers, Back King-street, Manchester. Also agent for
>
> **E. H. HAWKE & Co's PATENT SAFETY FUSE,**
>
> for blasting.
>
> Figure 151. *Manchester Courier & Lancashire General Advertiser*, **May 20 1848.**

Both of Hawke's businesses evidently prospered, because by 1851 he was employing 46 persons, and described himself as a 'rope, chain and safety fuse manufacturer'. He was still living in Tolgullow with his wife, three children and two servants.[3]

In the summer of 1851, Hawke exhibited specimens of his rope (but not fuse) in the Great Exhibition held in the Crystal Palace in London.[4] He was sufficiently prominent in the Gwennap area to be nominated in September 1851, together with other well-known mining men, to receive subscriptions for a presentation to the local landowner and mine investor, Michael Williams.

> At the Adventurers' Meeting at United Mines, on Friday last, it was proposed by Humphry Williams, Esq. M.P., that a TESTIMONIAL should be presented to MICHAEL WILLIAMS OF TREVINCE, Esq. by the Adventurers and Working Miners in the Gwennap District for his spirited and generous efforts to continue the working of these mines, under the most discouraging circumstances, in the early part of this year. This proposal was heartily and unanimously met by the Adventurers present, and a committee was appointed to carry it into effect.[5]

In December of that year, Edward Hawke attended the meeting and exhibition held by the Cornwall Agricultural Association in the Market House, Truro. Almost 40 men sat down to lunch at Pearce's Royal Hotel, where Hawke made a short speech after the toast to Queen Victoria, the Duke of Cornwall and Prince Albert. Hawke mentioned that his holding was tiny – 60 or 70 acres, and went on to say:- "it was only by the greatest application of the greatest industry and the greatest care and attention, that any man could hope to make a profit out of any business that he was acquainted with".[6]

A Fatal Explosion
On the 28th of July, 1855, an explosion in the fuse works in Tolgullow killed a

woman and a young girl. The explosion occurred at 4.25 pm, and it appeared that Mary Hawke (43) was killed when jumping out of a window, at which instant the whole building blew up and she was fatally injured under one of the collapsed walls. It is clear from reading the report of the inquest that Mary Hawke was jumping from an upstairs window – her means of escape down the stairs was blocked. The inquest reported that ten female workers were in the fuse works, and that two of them were working 'countering' the fuse. Some part of the fuse or countering twine became tangled in the machinery. The friction caused the ignition of some 'loose powder lying about the place', and this ignited two barrels containing about two hundredweight of gunpowder. Mary Hawke probably saw the loose powder ignite, and, realising that the barrels would explode with catastrophic results, immediately tried to escape via a window. Fanny Michell (13) was killed.

The newspaper reported that the premises were 'a heap of ruins and the shock was felt for some miles distant'. The inquest jury returned a verdict of 'Accidental Death'; and made no recommendations at all as to how safety might be improved.[7]

At some stage after this, Hawke resumed manufacturing his fuses in the Count House of nearby Wheal Unity.

An account in 1817 reported:- 'Wheal Unity is divided by the road leading to Chacewater from St. Day. About 30 years since, some poor men worked in this mine for the purpose of procuring tin, when they made such discoveries as led to the prosecution of the mine. Few under takings of a similar nature produced so much profit to the adventurers, for a series of years, as this has done'.[8] Spargo, writing in 1865, noted that Wheal Unity gave a profit of £360,000 and was idle.[9] A. K. Hamilton Jenkin, in his 1963 account of the Gwennap Mines, noted that Wheal Unity and its neighbour Poldice copper mine were amalgamated around 1800, although they continued to keep separate accounts.[10] In 1864 Wheal Unity, Poldice, Wheal Gorland, Wheal Maiden and the Carharrack Mine were combined as St Day United, but later became known as the Poldice Mines by 1870. However, Poldice and Wheal Unity were idle by 1864.[11]

A branch of the Great County Adit had been driven, by 1792, from Poldice, through Wheal Unity and in to Wheal Gorland. The water used in dressing the copper ore came from the adit of Pednandrea Mine in Redruth, carried via a leat of six miles in length. It passed through a tunnel over 4000 feet in length near Scorrier, delivering the water to the dressing floors of Wheal Unity, situated in the valley below Wheal Unity Wood.

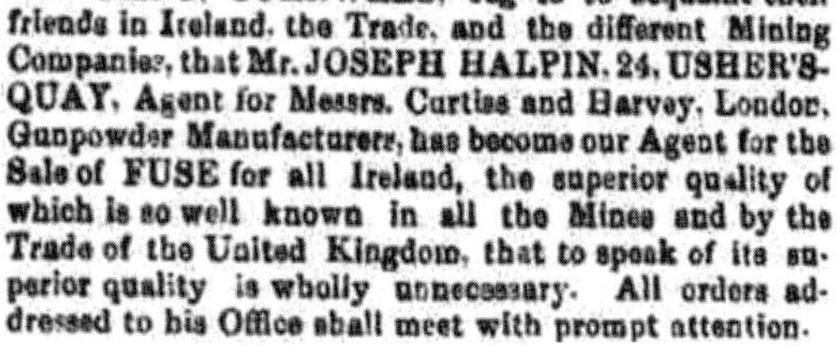

The Wheal Unity Count House was situated beside the Portreath to Poldice

THE UNITY PATENT SAFETY FUSE COMPANY, CORNWALL, beg to to acquaint their friends in Ireland, the Trade, and the different Mining Companies, that Mr. JOSEPH HALPIN, 24, USHER'S-QUAY, Agent for Messrs. Curtiss and Harvey, London, Gunpowder Manufacturers, has become our Agent for the Sale of FUSE for all Ireland, the superior quality of which is so well known in all the Mines and by the Trade of the United Kingdom, that to speak of its superior quality is wholly unnecessary. All orders addressed to his Office shall meet with prompt attention.

Figure 152. Advertisement in The *Freeman's Journal*, Dublin, 12th November, 1858

tramroad. This was constructed to link the copper mines around St. Day and Scorrier with Portreath harbour, on the north coast. It was in use as far as Scorrier House by 1812, and the section from Scorrier to Poldice was in use by early 1819. It was of 3' 6" gauge, and the wagons were pulled by horses throughout its length. It had almost fallen into disuse by 1865, and the tramplates were taken up and sold around 1882.[12]

By 1858, Hawke was not only manufacturing fuse, but exporting it to Ireland. The Cornish mine agents were active at this time in the mines of Cork, Waterford and Wicklow.

A sign of the general depression that was beginning to settle on the mining area of Gwennap was echoed in the report put out by the St. Day Institute in 1859. At their half-yearly meeting held in September 1859, the committee noted: - 'owing to so many young men having left the town for foreign countries, it has deprived the Institute of a great many members. Mr. E. H. Hawke, snr. and Mr. E. H. Hawke, jun., are the principal subscribers'.[13]

Edward Hawke senior was also a shareholder in the Cornwall Railway. The Chairman and Director was Dr George Smith, whose fuse company in Tuckingmill was a much larger competitor of Hawke.

Killifreth Mine
Edward Hawke and his son were also pursers to Killifreth copper mine, about a mile north of their house at Tregullow, near Tolgullow. Edward Hawke junior succeeded his father as Purser from 1863 to 1868. The mineral lord was Lord Falmouth. The mine re-opened in 1864, when Richard's shaft was renamed Hawke's shaft. The engine house, which can be seen today, was built in 1891 to house an 80-inch engine purchased from North Treskerby for £575. The mine closed in the 1890's, but again re-opened in 1912. An 85-inch engine was installed in Hawke's engine house, and, to give a better draught, the brickwork at the top of the stack was doubled in height.[14] This distinctive tall, slender stack can still be seen from the Scorrier to Chacewater road.

A Switch to Greenwich Mean Time
In September 1860 the inhabitants of St. Day and the surrounding area met in the Vestry Room to consider altering the present time to that of Greenwich. William Williams proposed, and E. H. Hawke seconded: - 'that for the future, Greenwich time be kept in this town and neighbourhood as at other places, with a view to meet the wishes of all interested in business matters'. The motion was unanimously passed.[15] By now Hawke employed 54 persons, and described himself in the census as 'manufacturer of chain, rope, safety fuse; and farming 82 acres'.[16] He was also a widower, his wife Charlotte having died in June 1860, at the age of 63. However, Edward, Hawke's only son, got married in September 1863 in Lyddington, Wiltshire.

His bride Emily was the daughter of William Wooldridge, R.N., Lieutenant of the Port of Gibraltar.[17] Their four children (one of whom died as a baby) were baptised at St. Day, when Edward described himself as a 'merchant'.

Edward Hawke junior died in June 1870, at the age of 39, from injuries sustained when he was thrown from his horse in Redruth.[18] The newspaper noted that he was 'united with his father in extensive rope spinning and safety fuse manufacture; he was the active representative of the business.' His funeral took place on January 19th 1870 in the parish church of St Day 'with Masonic Honours'.

Edward H. Hawke, senior, died in October 1871. The *Royal Cornwall Gazette* reported that he 'suddenly succumbed to gout in the stomach'.[19] He left two unmarried daughters, Julia (40) and Charlotte (41), who later left the district to reside in London.

Figure 153. Hawke's engine house, Killifreth. c.1920. Author's collection.

At the end of November there was an executors' sale of all the livestock, crops and farming implements on Hawke's farm, known as Tolgullow Farm, near Scorrier Gate station.[20] The following month his house, stables and grounds were put up for sale, together with houses he owned in Scorrier.[21]

In April 1873, the widow of Edward H. Hawke, junior, sold all her household furniture and effects by auction at the Polytechnic Hall, Falmouth.[22] She left the district with her children, and took up residence in London. She re-married at the end of 1880.

There is no account in the local newspapers at this time of the fuse works being for sale. The clerk of the rope works was Elisha Trewartha, who had been with Hawke's since 1841. He was one of Edward Hawke's (snr) executors, together with Hawke's two daughters and his solicitor W. J. Genn.

In November 1874, Hawke's executors put the rope manufacturing business up for sale by tender, together with a dwelling house and three cottages. There were extensive buildings and offices used for the manufacture of hemp and wire rope, and about seven acres of land.[23] It appeared that a partnership in the rope business was formed at this time between Trewartha, Genn, Julia and Charlotte Hawke, and Henry Backhouse Fox, a member of the very influential Quaker family from Falmouth. However, it seems that the rope business may not have found a purchaser. In October 1875, the partnership was dissolved.[24]

A FATAL FEBRUARY FIRE
DISASTROUS EXPLOSION AT ST. DAY. FIVE FEMALES BURNT TO DEATH

The Royal Cornwall Gazette carried this headline on Saturday February 27th, 1875, referring to a terrible explosion at Wheal Unity Safety Fuse Factory the previous Saturday.

> The factory, which is easily reached from Scorrier railway station, belongs to Sir. F. M. Williams. BART, MP, and therein employed about a dozen women and girls, none, we believe, married, and there is also a foreman, John Hamlin. The whole manufacture of the Safety Fuse was, therefore, carried on by one man as superintendent and several females.

The newspaper explained that Sir Frederick had only owned the factory for about two years and that previous to this it had been worked by 'Mr. Fox'.

The explosion took place at about 11am, when, according to the newspaper, the most dangerous part of the manufacturing process was taking place – 'laying gunpowder on the tapes. To do this a little wagon filled with powder is made to run on a small tramway and emits a certain amount of powder as it travels along'.

All five of the women who were killed were working upstairs in the old Count House, reached by a single staircase. The women who were killed almost instantly were:- Anne Davey (40), Christiana Michell (16), Elizabeth Anne Pooley (15) and Eliza Jane James (34). Margaretta Long (17) died the next day. Long could not give the cause of the explosion. She saw the three young girls rush madly to the windows. She herself knocked out the glass from one of the windows and jumped out.

The women working downstairs were also burnt and injured. They were named as Elizabeth Anne Davey, Susan James, Kate Luke, Cecilia Stephens and 'a girl named Martin'. It seemed that both Davey women had been injured in the explosion in 1855 at the old factory in Tolgullow.

Hundreds of people quickly made their way to the factory. Rescue was impossible. The newspaper reported that 'a short time after the explosion the factory, which was a small but compactly constructed building, was enveloped in one mass of flame.' It was thought that the flames spread so rapidly because jute caught fire. This flammable material used in the manufacture of fuses was stored in the factory. The newspaper noted that the quantity of gunpowder in the factory was very small, as the walls were standing and the windows were not entirely blown out. A 'gunpowder

–proof' house which stood next to the factory contained over one hundred-weight of gunpowder.

The Royal Cornwall Gazette gave a very full account of the tragedy:

A very large number of miners from West Poldice, which is about 200 or 300 yards from the factory, under Captain Joseph Cock, poured water on the flames, and it was through their efforts, and those of PC Drewitt, of St. Day, that the bodies, or remains, of the unfortunate females were found, quicker than otherwise they would be. One man in particular made a strenuous effort to save his child. Elijah Pooley, when he heard of the explosion, rushed with all possible speed to the factory, went into the burning building at the greater risk of his life, but the distracted father and courageous man engaged in a hopeless effort. He was burnt very much, but he lent great assistance to others. Capt. Mayne, Unity Wood, and an engine man named John Harris, behaved in a most courageous manner. At the time of the explosion the boiler, which is built in a house adjoining the factory, was in full working order, and if the flames had reached this house it can easily be imagined that another very disastrous explosion would have occurred. To prevent this, Capt. Mayne and Harris, at a great risk to their lives, for there was a strong possibility that they would have been shattered to pieces, succeeded in getting a rope round the safety valve, then threw the rope over a beam, and leaving the boiler house, pulled the valve from a distance and so prevented a terrible explosion.'

The charred remains of the four victims were placed into an outhouse, to await the inquest. Sir Frederick Williams arrived at the scene of the fire and was 'deeply affected'.

The Inquest
The inquest was held on the Monday following the fire, in Bennett's Commercial Hotel, St. Day. Sir Frederick Williams was present, and accompanied Mr Carlyon, the Coroner, and the jury to view the premises, the remains of the deceased, and the body of Margaretta Long.

At the inquest, Rosella Stephens stated that at the time of the explosion she had been working downstairs 'taping', and she made her way through wide-open doors into the yard. She had worked for the Company for 16 years. She testified that slippers, provided by the Company, were to be worn at all times. She denied that all the doors were locked, and added 'when the women went into the yard for anything, they changed their slippers'. Sir Frederick Williams said the only doors which were kept shut were those leading to the yard, which were 'fastened in order to keep out the public who had no business near the works'.

Caroline Jones, who had been working downstairs at her machine, was enveloped in dark smoke. She crawled towards the door and made her escape. Catherine Luke was downstairs also, at the tape machine. She thought she saw 'fire pass before her eyes' and found herself in darkness. When she recovered, she found herself in another part of the room. There was a hole cut into one of the screens near her; she crawled through this into the fuse-room and then into the yard. She thought there might have been three or four pounds of gunpowder in the upper room.

John Pooley (sic), whose daughter was killed, testified that he rushed to the scene, and saw one girl jump over the outer wall, and fall to the road. He caught four more girls who were on top of the outer wall. He himself got over the big doors into the yard, where he saw John Hamlin, the foreman, and the girl who worked the engine putting out the flames on the clothes on Margaretta Long. She managed to walk into the road. Hamlin advised him to leave, because there was a small quantity of gunpowder on the premises. This did, in fact, explode. He could not see any of the girls, and it was impossible to get upstairs.

A surgeon, W. P. Hodge testified that he treated Margaretta Long at her parents' house. Her clothes were taken off, and she was treated with oil and lint. He said that she was badly burnt over her back, shoulders and chest. She told him that the first thing she heard was an explosion. She died at midnight on Sunday, the day after the fire.

The inquest was adjourned for a fortnight to enable John Hamlin time to recover enough to give evidence.

Sir Frederick Williams asked if his agent James Tregonning could be questioned when the inquest resumed. He wanted to prove that the recommendations of the Government Inspector for the safety of the works had been carried out. He stated: - "I think this would be satisfactory to the relatives of the deceased, to show that every precaution had been taken. I am sure, for my part, I have always desired to make everything safe for my workpeople."

The five women were buried in a communal grave four days later, with all funeral expenses paid by Sir Frederick.

The Adjourned Inquest
The inquest resumed on Monday, March 8th. Major Ford, R.A., one of the Government's Inspectors of Gunpowder Works, was present at the inquiry.[25]

John Hamlin, the foreman, testified that he was in the engine-room at the time the fire occurred. He heard no explosion. Something struck him on the head. He got caught in the engine, but when he came to, he could not see anything – it was as if he had been blinded. His sight returned, and he went in search of the girls. He saw Margaretta Long at a window. He went for a ladder, but she jumped before he got back. None of the windows were locked. He thought that there were between 24 and

30 pounds of gunpowder in the room, in six to eight parcels. It was loose gunpowder, and fire might easily have spread between them if it exploded. Hamlin stated that he could not account for the explosion – all the persons working in the upper room being dead.

In answer to the Coroner, Hamlin said that in the morning Long was working in a shed in the yard. He saw her boots, which were off. He presumed she was wearing slippers. If she was wearing slippers he presumed she went into the (fuse) room from the yard with the slippers on. It was against the rules to wear the slippers in the yard, and walk from there into the room wearing the slippers.

Major Ford asked Hamlin to produce a copy of the printed regulations in force at the factory. Hamlin replied that this was not possible – they were not printed. He said that the women 'perfectly understood the regulations'. They were to take their boots off at the door of the upper room, and to put on their slippers. He had told them that if they did not obey this order, they would be discharged.

There followed a series of very penetrating questions to Hamlin from both Major Ford and the Coroner.

Q. (Major Ford) 'Did the women wear any pockets?'
A. (Hamlin) 'I gave orders for them not to wear any, and they told me they did not.'
Q. 'As a matter of fact, do you know they had no pockets?'
A. 'No, I do not; never examined one of them to see.'
Q. 'How did you remove the semi-manufactured fuse from the room?'
A. 'It was always carried away by hand, covered over by a piece of hessian. It was removed about every other day. It was kept at one end of the room, partitioned off from the rest.'
Q. 'Do you know how much was in the room at the time of the explosion?'
A. 'I cannot tell. Somewhere about 300 coils, a coil being twenty four feet.'
Q. 'When did you last remove the semi-manufactured fuse from the room?'
A. 'On the afternoon before – the Friday.'
Q. 'What are your regulations for sweeping the floor?'
A. 'It is swept every night, half an hour being devoted to cleaning up the powder on the floor. Every girl had to clean her place thoroughly.
Q. 'Where are the sweepings placed?'
A. 'In a covered barrel.'
Q. (From the Coroner). 'Is it the place of anyone to see the boots taken off and the slippers put on?'
A. 'There were no regulations regarding that. If I saw anyone who was not attending

to the orders, she would be fined for the first offence, and discharged for the second.'

Q. (The Coroner). 'Did you ever detect anyone going into the room with boots on?'

A. 'No, they would be afraid to let me see them.'

Q. 'Did you ever hear that such a thing was done?'

A. 'I have no doubt it was done. Such things are done in every place. There was never a place yet where the rules were not broken.'

Q. 'Have you anyone to carry out the rules when you were absent?'

A. 'There was no one there that I could trust when I was away. The machines often got out of order; there was always something wanting repairs, and when this was so, the machines were taken to pieces, and the work done in the engine room.'

James Tregonning testified next. He said he was the agent for Sir F. M. Williams, Bart., M.P., and as such was connected with the Unity Safety Fuse Works. It was his duty to go to the factory to see that all was right, and to ask if there were any complaints to be made. He went once a week for this purpose, and once a fortnight to pay the people. It was not his duty to see that the regulations of the factory were carried out; that work belonged to Hamlin.

Mr. Tregonning added that he had a letter in his hand which he desired to have read, as it had been sent from Major Majendie, Inspector of Gunpowder Works, in reference to improvements made in the factory since it had come into the possession of Sir. F. M. Williams. The letter stated:-

On inspecting your factory on the 17th March, 1874, in company with Major Ford, I noted with satisfaction the very marked and considerable improvement which had been effected, and the ready adoption of the majority of the suggestions which I deemed it my duty to make on the occasion of my last inspection in April 1873.

You will permit me to call your attention to such points as still appear to me to need improvement or alteration.

1. The sweepings in the fuse-room should be kept in a closed vessel, instead of a bucket, and more care should be taken to prevent the spilling of powder on to the floor. I found in the portion of the room where the barrel of sweepings was situated a considerable quantity of powder spilt and lying on the crevices of the uncovered part of the floor.
2. It would be prudent to adopt measures to prevent the made fuses from accumulating in the fuse-making room after manufacture, so that in the event of an accident in that room, the consequences would be at a minimum.

3. The walls of the fuse-making room behind and beside the machines should be wood skirted to a height of three or four feet, and, if practicable, I would strongly recommend that this portion of the room be screened off from the remainder (care being taken to leave ready and sufficient means of exit for every woman in the room.)
4. The window of the 'dry' should be externally protected with wire.
5. A code of working rules should be drawn up, a copy placed in every building and given to each work-person affected.

Agent James Tregonning said that the suggested improvements had been made, and that at the time of the accident the rules, which had been drawn up, were being printed. He said 300 coils of semi-manufactured fuse was not so much to leave in the room; they made 600 coils a day, so it represented only half a day's work.

Major Ford, the Government Inspector, said it was desirable to remove the semi-manufactured fuse as often as possible, to prevent it smouldering if a fire broke out, and thus causing suffocation. Hamlin replied: - "Then it should be removed twice a day, sir." Major Ford said: -"Oftener, if possible".

A surgeon, George Mitchell was called to give evidence. He testified that he went to Margaretta Long's home after the fire, and assisted Mr. Hodge in taking off her clothes and dressing her wounds. On her feet she had a pair of old leather boots. In one, he observed three or four iron 'sprigs' or small nails.

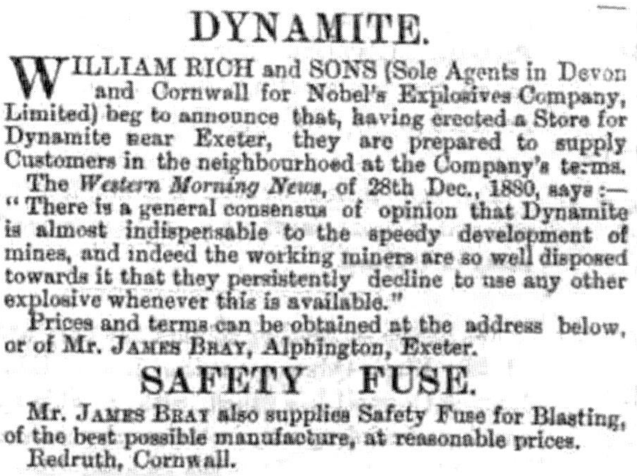

The Coroner said there was no doubt that Long infringed the rules, and probably that was the cause of the explosion. They knew of no other cause. He did not think any blame could be attached to anyone, because, if the rules were carried out, the explosion would probably not have occurred. Although the rules had not yet been printed, there was no doubt that all who worked in the factory were well acquainted with them.

The jury found 'that the four women not identified died from suffocation, and that Long died from burns and shock to the nervous system.'

Figure 154. *The Western Times*. 10th March 1881

The Explosives Act, 1875

This Act came into force on the 1st of January 1876, and strictly regulated the conditions under which factories manufacturing gunpowder (which included fuses) could operate. The gunpowder magazines were also tightly regulated. Licences had to be obtained for both.

In July 1876 the Unity Patent Safety Fuse Company applied for the licence for their factory at Wheal Unity. The application came from the Tregullow Offices of the William's estate, and was almost certainly made by James Tregonning, the William's agent. He also applied for a licence for a gunpowder magazine situated in the nearby Wheal Gorland.[26]

The Death of Sir Frederick Williams

On Tuesday September 3rd, 1878, Sir Frederick Williams died suddenly at the age of 78. He was in Devon, visiting part of his family's estate at Heanton Court.

The fuse factory appeared to be in decline. In 1881, Elijah Trewartha (66), Edward Hawke's clerk in the old rope works, was running the Company. He was listed in the census as a manufacturer of safety fuse employing one man and five females.

Around this time the fuse works were purchased by William Rich, a prominent mine agent from Pednandrea in Redruth. He was the agent in Cornwall for British Dynamite (which became Nobel's Explosives Ltd. in 1876).[27] He would have been aware that the Unity fuse works may not have been prospering. He put his son-in-law William Bennetts (33) in charge. Bennetts had been working in Swansea as a mercantile clerk – his only qualification for the position was as husband to Rich's daughter Mary. Bennetts (a native of St. Cleer, near Liskeard) arrived in Gwennap with his wife and one year old son Samuel, and took up residence at Pink Moors Spring. Four more children quickly arrived – Frederick William, Rosa, Mary Eveline and Ralph.

By an agreement dated June 24th 1884, William Rich had leased, for a term of 60 years:-

> All that piece of land near Little Beside and adjoining Wheal Unity Accounting House in the parish of Gwennap …………..formerly in the occupation of James Mitchell but then of the said William Rich being part of the tenement known by the name of Shilstones and parcel of the manor of Tolgullow otherwise St Day, which said premises are marked on the map of the said Manor with the letters N.N. together with the Accounting House erected thereon'.[28]

The original lease for the plot of ground had been signed in October 1842. It was for: - 'a plot of land near Little Beside and adjoining Wheal Unity accounting house built there'. It was leased to James Michell of Gwennap, innkeeper, for 99 years. It is not known when the Count House was first built. The mine was first worked about 1791, so the Count House could have been over fifty years old when the fuse

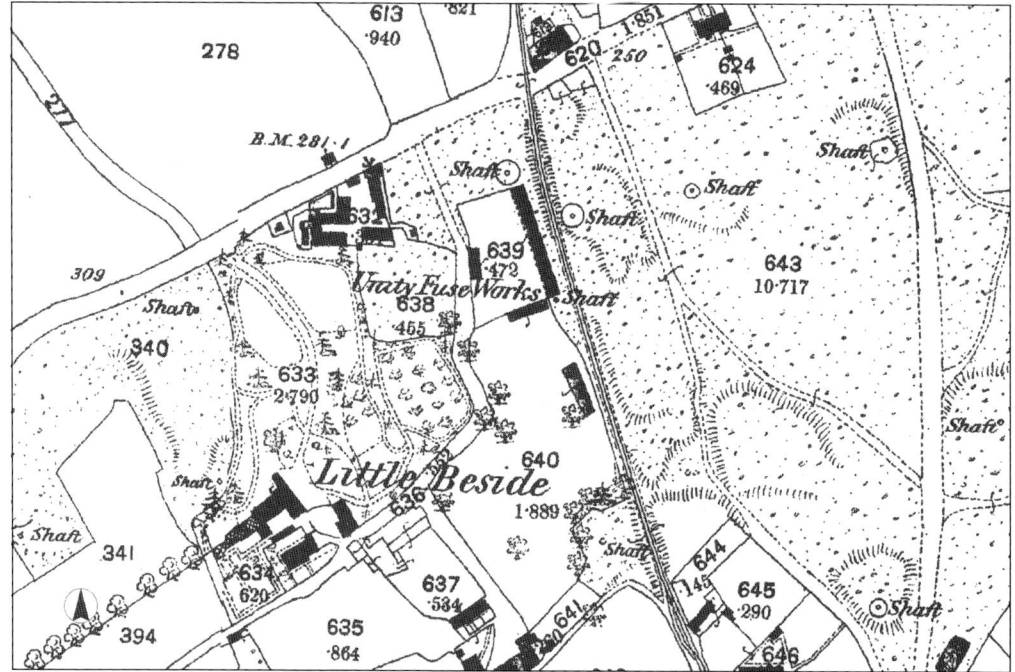

Figure 155. 1886 map of Little Beside and the Unity Fuse Works.

Company took it over around 1855.

The 1886 Ordnance Survey map shows the Fuse Works beside the Poldice to Portreath tramway. The long N to S building was probably where the fuses were laid out in length for varnishing.

In October 1892 the Unity Safety Fuse Company leased more land from Charles Dysart Teague and Richard Henry Teague, for a term of 52 years from 24 June 1892.

> Plot of land marked off out of a close, recently improved from Croft part of Little Beside or St Day Manor.[29]

It appeared that the fuse works were enlarged at this time, and that may have been the reason for leasing more land. In December 1892, under the Explosives Act and the Local Government Act of 1888, the new owners of the 'Unity Safety Fuse Company of Scorrier' applied to the County Council for their assent to the establishment of a new factory or magazine at Wheal Unity Old Mine.[30]

In July 1898, William Rich died at the age of 78. His two sons Joseph and Harry inherited the fuse works and made William Bennetts a partner in the concern.

In 1906 the two Rich brothers and Bennetts took two more partners into the concern, namely Sir George J. Smith and his son G. E. S. Smith. The Smiths were owners of Bickford, Smith and Co. Ltd in Tuckingmill. They purchased two-thirds of the Unity Fuse Company, in two equal one-third parts, and paid £1,333 6s 8d. This included the premises, and an agreement that the terms of the original leases would be observed.[31]

The employees of the fuse works had an outing in August 1907. They travelled by train to Padstow. Sir George Smith had always given his employees in Tuckingmill an annual outing, and this trip was likely to have been suggested by him.[32]

The fuse works were not doing well, and a number of women were discharged in May 1908. *The West Briton* reported:- 'this is the second discharge recently made and is causing great distress amongst the work girls'.[33] In October 1908, the Unity Fuse Works were given a rateable value of £120. This was £20 more than the value given to the St. Day Fire Brick and Clay Company, two miles away on the south western edge of St. Day.[34]

On the 28th of August 1911 another agreement was signed between the Rich brothers, William Bennetts and the Smiths. In the original 1906 agreement it had been agreed that Bennetts and the Rich brothers could, if they wished, opt out of the partnership after five years. They all wished to do this, and were therefore entitled to their one-third share of the business, divided into three equal parts. Therefore the agreed sum of £733 6s 8d was divided into three and shared equally between them.[35]

William Bennetts was 62 years of age in 1911, and was living in Tolgullow. His wife Mary died in 1916, at the age of 63, when she was described as 'the wife of Captain Bennetts, of the Unity Safety Fuse Works'.[36] The Rich brothers were merchants in Redruth.

Yet another agreement was signed on the 24th of April, 1917, this time between Sir George J. Smith and G. E. S. Smith, and Bickford, Smith and Company Limited.. The two Smiths had been acting as Trustees for Bickford, Smith and Co. Ltd when they had purchased the Unity Safety Fuse Company, the money for which had, in fact, been paid by Bickford, Smith and Co. Ltd. This Company now wanted to become the legal owners, and Sir George J. Smith and G. E. S. Smith assigned the Unity Fuse Company, its leases and premises to Bickford, Smith and Co. Ltd.[37]

Bickford, Smith and Co. Ltd did not close the Unity Fuse Company until after the 1914-1918 War. In 1918, Sir Harry McGowan organised the merger of most of the explosive companies still in operation (see Chapter 9). Bickford's put the Unity Fuse Company into the scheme. Sir George J. Smith became the deputy-chairman of the new Explosives Trade Ltd. The exchange value of shares in companies such as Bickford, Smith and Co. Ltd (of which the Unity fuse factory was now a part), was not made public, as the shares were owned by family members and not traded openly.

The grandsons of Edward Henry Hawke, the founder of the Unity Fuse Works, rose to prominence in London. John Anthony Hawke became a King's Counsel, and a judge, and received a Knighthood from King George V in 1928. Edward Drummond Hay Hawke became an eminent physician. He died in 1928, two weeks after having a golfing holiday in Cornwall, when he stayed at the Porthminister Hotel, St. Ives.[38] His brother had formerly represented this town as an M.P. in the 1920s.

William Bennetts continued living in Tolgullow, and died in April 1930, at the age of 83.[39] He is buried in Redruth cemetery, on the road to St Day.

References, Chapter 14

1. *Royal Cornwall Gazette*, October 24, 1829
2. Cornwall Online Parish Clerk.
3. 1851 Census, Gwennap, St Day. District 2b
4. *Royal Cornwall Gazette*, May 9, 1851
5. *Royal Cornwall Gazette*, September 26, 1851
6. *Royal Cornwall Gazette*, December 26, 1851
7. *Royal Cornwall Gazette*, August 3, 1855
8. *The Gazetteer of the County of Cornwall, 1817.* Online
9. Spargo, Thomas. *The Mines of Cornwall*. Part IV. 1865
10. Jenkin, A. K. Hamilton. *Mines and Miners of Cornwall*. Part VI. 1963
11. Collins, J. H. *Observations of the West of England Mining Region*. 1912
12. Fairclough, A. *The Story of Cornwall's Railways*. Tor Mark Press. 1970
13. *Royal Cornwall Gazette*, September 28, 1859
14. Buckley, Allen. *Killifreth Mine*. Penhellick Publications. 2011
15. *Royal Cornwall Gazette*, September 14, 1860
16. 1861 Census
17. *Devizes and Wiltshire Gazette*, September 17, 1863
18. *Royal Cornwall Gazette*, January 15, 1870
19. *Royal Cornwall Gazette*, October 21, 1871
20. *Royal Cornwall Gazette*, November 16, 1871
21. *Royal Cornwall Gazette*, December 16, 1871
22. *The West Briton*, April 17, 1873
23. *The West Briton*, November 26, 1874
24. *The London Gazette*, October 26, 1875
25. *Royal Cornwall Gazette*, March 13, 1875
26. *Royal Cornwall Gazette*, July 22, 1876
27. *Falmouth Packet*, April 29, 1876
28. County Record Office, WH/1/4759
29. County Record Office, WH/1/4759

30. *Royal Cornwall Gazette*, December 15, 1892
31. County Record Office, WH/1/4579
32. *The Cornishman*, August 15, 1907
33. *The West Briton*, May 21, 1908
34. *The West Briton*, October 22, 1908
35. County Record Office, WH/1/4759
36. *The West Briton*, May 4, 1916
37. County Record Office, WH/1/4759
38. *The Cornishman*, September 6, 1928
39. *The Cornishman*, April 10, 1930

CHAPTER 15

Edward Tangye's Fuse Works, Redruth
1886-1899

Edward Tangye was one of the Tangye brothers who started the famous Cornwall Engineering Works in Birmingham in the early 1860s. In the 1851 census he was living in Broad Lane, Illogan, aged 16, with his Quaker parents, his brother James (25), Joseph (24), a younger brother George (15) and his sister Sarah (6). He was described as a 'farmer of 12 acres, employing 1 labourer.' His older brothers were working nearby in the Penhellick fuse factory of William Brunton. By 1861, he and his brother George had moved to Birmingham to join Joseph and James, and another brother, Richard.

He left Birmingham in 1880, and returned to Cornwall with his wife Anne and small son Ernest. His son Alfred was born in Redruth in 1880, where the family lived in Trewirgie. He described himself in the census as a 'mechanical engineer, retired'.[1] He was 37 years old, and it appeared that ill-health had brought about his retirement.[2]

By 1881, he had moved to the northern part of the town. His house was called 'The Elms' and it was located at the junction of Green Lane and Drump Road. He had probably moved there to accommodate his growing family, which by now numbered eight children. In the census, he described himself as a 'Mechanical Engineer'. The British and Foreign Safety Fuse Factory was not far away from his house, on the western side of the Plain-an-Gwarry area.

In March 1885 Edward Tangye took out a patent for a tube-rolling machine.[3] He further patented his invention in France and in the United States.[4] In December 1886, he applied to the Justices of the Petty Sessional Division, sitting in Camborne, for their assent to the establishment of an explosives factory at The Elms.[5] The next week, the newspaper reported that as there was no opposition, the application to build a new safety fuse works was granted.[6]

Tangye's Patented Fuse, 1885
His specification stated: -

My invention relates to improvements in the manufacture of safety damp-proof metallic fuse, and the objects of my improvements are, first, to reduce the thickness of the metal coating, and consequently the weight of the fuse to a minimum, and, secondly, to provide an article which shall be absolutely free from imperfections and perfectly damp-proof.

I am aware that metal-coated fuses have been in use prior to this invention, and therefore do not claim the use of metal, broadly, as a case for the powder, but what I do claim as my invention and desire to secure by Letters Patent, is -

1. The method of making metallic-coated fuse, which consists in first loading a short tube with gunpowder or other suitable material and closing the ends of it, then passing the said loaded tube between two rolls having a series of meeting grooves, forming a gradually- reducing series of alternately circular and oval holes, which elongate and reduce the thickness of the said loaded tube, and finally passing the said tube through grooves which change its sectional form without elongating it, and thereby granulate its contents, substantially as set forth.

2. As an article of manufacture, a rolled fuse consisting of a core of inflammable material having a very thin enclosing metallic case formed about it, substantially as set forth.

The Elms had an extremely large back garden, bounded on two sides by public roads. It was here that Tangye erected his buildings and began the manufacture of his metallic fuses. In 1891, he was living there with his wife Anne (44) and eleven children ranging in age from 20 year old Alfred to 5 year old Lancelot.[7]

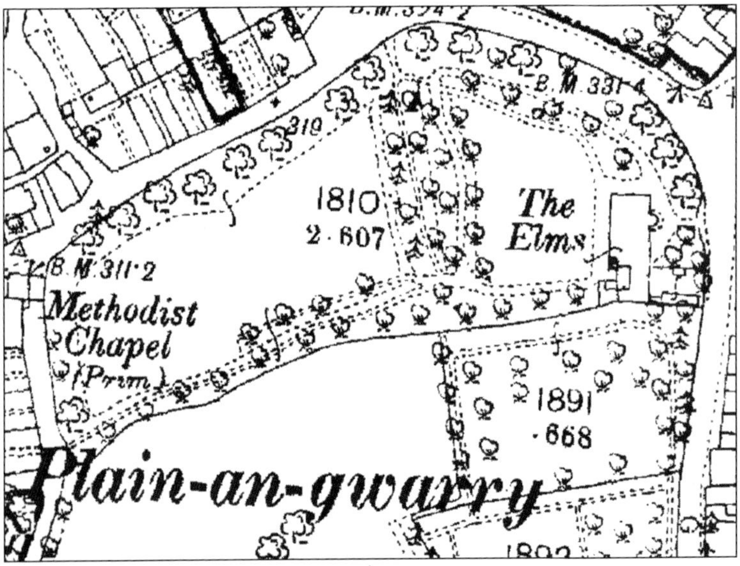

He registered his Trade Mark in June 1888.[8] The Trade Mark shows a column of water blasted from the sea. Tangye clearly wanted the waterproof qualities of his fuse to be recognised. He gave his fuses the trade name of 'Excelsior' and advertised a

Figure 156. Map of 'The Elms' in 1880, before Edward Tangye built his fuse works.

204

Figure 157. Advertisement for Tangye's fuses, 1898.

blue thread running throughout the length of the fuse.

The Fuse Works is Destroyed by Fire

On Thursday evening, August 10th, 1893, Tangye's fuse works caught fire. The blaze started in the pitch-house attached to the factory. Other nearby engineering buildings were in the grounds, and the blaze spread to them, destroying everything apart from the brickwork. The works contained a five-horsepower engine, machinery for making the fuse, turning lathes, planing machines and other appliances.

Small explosions of gunpowder and the fearful anticipation that the engine boiler would explode, kept onlookers at a distance. However, steam was released, and the boiler was saved. Fuse and tools were rescued, as far as that was possible.

Vineries Destroyed

The Royal Cornwall Gazette reported:-

Much sympathy is expressed with Mr. Edward Tangye, not only for the pecuniary loss sustained......... but also for the way in which his pleasant grounds and extensive vineries have been destroyed. Through some mistake on the night of the fire, some person, who evidently thought the vineries would be burnt down with the fuse manufactory, suggested that those who were working might as well go into the vineries and bring out the fruit, refreshing themselves at the same time. This suggestion was immediately and cheerfully followed by a number of workers and onlookers, who speedily appropriated a large quantity of the crop, in fact, some hundredweight of grapes, until the police interfered and cleared the house. The fruit on the trees in the garden were also laid under contribution.[9]

His older brother James offered to help with machinery from his workshop in nearby Illogan He had returned to Cornwall on his retirement, and had built a small engineering workshop at his house, known as 'Aviary Cottage'. The Tangye Bros. Foundry in Birmingham also offered assistance.[10]

Two weeks later Edward Tangye submitted plans for new buildings to replace those which had burnt down. The Redruth Local Board decided to visit the site. He appeared to have replaced the buildings.[11] The 1908 OS map for the area shows two long buildings in the grounds behind the main house.

Figure 158. 1906 map showing a building and greenhousess to the rear of The Elms

Figure 159. 1924. The Elms (right), with two long buildings to the rear.

By 1899, however, the fuse works were closed. Nobel's Explosives Company had bought the machinery in 1899, and moved it to their Swansea Fuse Factory.[12] Tangye himself left Redruth around this time too, and returned to Birmingham shortly before the death of his son Alfred in February 1900.[13] He took up residence in the Manor House in Knowle with his wife and seven of his children.[14] He died there on December 9, 1909, aged 76.

The Cornishman reported his death, and added that at the Manor House: - '……..he resumed his horticultural work, giving special attention to the culture of apples, in which he was remarkably successful'.[15]

References, Chapter 15

1. 1881 census
2. *The Cornishman*, December 16, 1909
3. *Birmingham Post*, April 9, 1885
4. Patent 173,458 (France). Patent 364318 USA. Online. www.google.com/patents/US364318
5. *Royal Cornwall Gazette*, December 17, 1886
6. *Royal Cornwall Gazette*, December 31, 1886
7. 1891 census
8. *Western Times*, June 25, 1888
9. *Royal Cornwall Gazette*, August 17, 1893
10. *The Cornishman*, August 24, 1893
11. *Royal Cornwall Gazette*, August 31, 1893
12. *The Cornishman*, December 16, 1909. Also see Ref. 1. Chap 3. G. L. Wilson.
13. *The West Briton*, February 15, 1900
14. 1901 census. Birmingham
15. *The Cornishman*, December 16, 1909

PART 4

A CARADON AREA FUSE WORKS

CHAPTER 16

The Tremar Coombe Fuse Works
c.1855-1870

In October 1859, the *London Gazette* carried a notice to report that Joseph Opie, a fuse manufacturer, had submitted his invention of 'improvements in instruments or apparatus for charging holes in blasting operations, parts of which are also applicable for like purposes'. His address was 'Tremar, Cornwall'.[1]

The fuse works were in Tremar Coombe, a small village two miles north of Liskeard, and just to the south of Bodmin Moor and the mines around Minions. The development of copper mines around Caradon Hill led to the expansion of what was then a hamlet. In 1855, South Caradon copper mine employed 600 people. West Caradon employed over 350 men. Other mines were Gonamena, New West Caradon, East Wheal Agar and Craddock Moor. A United Methodist Chapel was built in 1863 to serve the mine workers in Tremar Coombe.

In 1841, Joseph Opie, aged 16, was living in Stithians with his mother and eight siblings. There were also six other people living in the house. His brother Thomas (16) was a copper miner, and his sisters Susanna (14) and Elizabeth (13) were also employed as mine workers. He again appears in the records in 1851, living as a lodger in Halsetown, near St Ives. He gave his occupation as 'safety fuse manufacturer'. William Thomas (63) was also living in Halsetown, where he described himself as a 'rope manufacturer employing six men'. His son James Henry Thomas (26), who was living with his parents, described himself as a 'Rope and Safety Fuse Manufacturer'.[2] It seems that the Thomases were making a small quantity of safety fuse in their ropeworks, and supplying it to the local mines. The mines here were tin mines, the largest being St. Ives Consols and Rosewall Hill. It is likely that Joseph Opie was employed here making fuses and decided to strike out on his own.

He moved to Tremar Coombe sometime after this, and set up a small fuse works there. In 1861 he was lodging in Tremar Coombe, in Middle Coombe, with Thomas Jones, a copper miner, and his wife. He gave his occupation in the census as 'safety fuse maker'.[3]

Also living there in Higher Coombe, was Thomas Opie (37), from Stithians, who gave his occupation as 'copper miner and grocer'. He was probably Joseph's brother. Grace Opie (47) was also living in Tremar Coombe with her daughter Mary (17), who was working in Joseph's fuse works. Her husband was Richard Opie from Stithians, who was almost certainly another brother of Joseph Opie.

Application for a Licence

The Cornwall Epiphany sessions were held at the County Hall, Bodmin, on Tuesday January 1st 1861. On Thursday January 3rd, an application was made to the Court for a licence to establish a safety fuse manufactory in Tremar Coombe. The applicants were Richard Hawke, of Liskeard, a merchant; Joseph Medland of Liskeard, an accountant; and Mr Opie of Tremar Coombe. In the course of the application it transpired that Opie had been making fuses in the parish of St Cleer 'for some time'.[4] Hawke, Medland and Opie applied for a licence to establish a new factory for the making and sale of safety fuse. The application stated: - 'the site of the proposed manufactory was admirably situated, being at a safe distance from any dwelling house, and no damage to property or life was consequently to be apprehended.'

At the hearing, Henry Rice, an architect and surveyor, produced a plan of Tremar Coombe, showing the site of the proposed works and the neighbouring cottages. He noted: -

- The nearest cottage was distant some 50 yards from the spot, and the next was 87 yards distant.
- The spot was a proper and suitable one for such a manufactory.
- In his opinion it was the best that could be found in the neighbourhood.
- There was no danger to the neighbourhood.
- Mr. Morris, the owner of the property, consented to the erection of the fuse manufactory.
- It was in the centre of an extensive mining district.
- The nearest cottages were from 10 to 15 feet higher than the site of the proposed building and the next (Probert's cottage) was 30 feet higher.
- The building would be about 60 feet long. This was the same as the present manufactory.

Rice stated that the reason Mr. Morris had granted permission was the fact that the manufacture of safety fuses had been carried on for six years in the parish, in the middle of a number of houses, without any accident occurring.

Joseph Opie stated that he had been manufacturing safety fuse for nearly 7 years at Tremar Coombe, and wanted a more convenient building. He estimated that between 1,400 and 1,600 coils of fuse a month would be made in the new building. He said the yarn used came from a manufactory and the gunpowder came from Herodsfoot.[5] He felt that no more than a hundredweight (112 pounds) of gunpowder would be on the premises at any one time.

On being asked what amount of gunpowder would be required at any time in the factory, Opie replied that about one hundredweight would be ample to meet every requirement, and the Court might limit it to that amount.

One of the magistrates asked if a hundredweight of gunpowder exploded, would it injure the cottages nearest to it. Henry Rice replied that he thought the weakest part of the factory would give way first and that it would be blown up into the air. The Court granted the application, limiting the amount of gunpowder to be kept in the factory to one hundredweight, and stipulated that it should only be used to make safety fuse.

Figure 160. 1883 map. The fuse factory is south of the Chapel, on plot number 1312.

Richard Hawke was born in Helston in 1823 and became a barber in the Caradon mines. He invested in the mines, and by the early 1850s established himself as a share broker, selling shares in the Cornwall and Devon mines.

MR. RICHARD HAWKE, MINE SHARE BROKER, LISKEARD, CORNWALL.

Figure 161. Advertisement from *The Mining Journal*, August 20, 1853

He had interests, amongst others, in South Caradon Mine, Mineral Bottom Mine, Wheal Norris, and West Chiverton. The latter lead mine east of St Agnes seemed to have provided the bulk of his wealth. By 1870, he was taking annual dividends from this mine of over £12,000 per year.[6] Hawke would have known how much money the mines around Liskeard were paying for fuse. He may have thought that a small fuse works, manufacturing fuses for all the local mines, would be a profitable business enterprise.

During the 1860s, all the Cornish copper mines were hit by a slump in the price

Figure 162. The fuse works building can be seen immediately to the right of the Chapel. c.1920. CRO

of copper, and the mines around Tremar Coombe were no different. The small mines closed and South Caradon, although it did not close, was badly affected. The fuse factory closed also; there is no mention of it in the 1871 census.

Richard Hawke died in November 1887. He held the lease, granted by the Prince of Wales, on South Phoenix Mine, Minions, with Richard Gundry and Henry Houseman. A new company was formed after his death. This company was to be worked on the cost-book principle, and 12,000 shares were to be issued, half of which were to go to Thomas and William Gundry, Houseman, N. J. Ridley, A. C. L. Glubb and Hawke's widow Sarah.[7]

The Cornishman newspaper noted in January 1888: - 'the late Mr. Richard Hawke of Liskeard, once a Helston barber, left all his property - £135,000 - absolutely to his wife Sarah'.[8]

'CREMATION AND FUNERAL OF A CORNISHMAN'
Richard Hawke's cremation and funeral made headlines in newspapers across the south and west of the country. *The Hampshire Advertiser* carried the following headline: - 'Cremation and Funeral of a Cornishman', and carried the following report:-

A funeral of a very unusual character took place on Monday in Liskeard. The body of Mr. Richard Hawke of Westbourne, Liskeard, was, at his own special request, cremated. Last Thursday, encased in a thin coffin, the body was taken to Woking, where there is an establishment for the burning of dead bodies...... The ashes were brought back to Liskeard to be interred.

The interment of the ashes took place on Monday in the grounds of Westbourne, the residence of Mr. Hawke at Liskeard. About a hundred invitations had been issued upon gilt-edged cards and there was a large attendance.[9]

Mr. Hingston, the family surgeon, conducted a small service, and Hawke's ashes were interred in a small brick vault in the garden.

Grace Opie, whose daughter Mary once worked in the factory, was living in Chapel Row in 1881, where she was described as a 'pauper'. James Opie (58) was living on his own a few doors away, next to Thomas Opie (32), a tin mine labourer. Opie was described as a 'General Labourer'.[9] He died in December 1881.

The Fuse Works

Half of the former fuse works on Chapel Row has been demolished to make way for a space to park cars. It had been used to keep pigs. The wall bordering Chapel Row is still largely intact, although one end appears to have been lowered. A door into the factory has been blocked with a wooden door, although the step can still be seen. The roof is corrugated iron. The interior yard is overgrown with brambles. A round opening is visible in the wall, where a gate across the lane was once in place.

References, Chapter 16

1. *The London Gazette*, October 7, 1859
2. 1851 census
3. 1861 census
4. *Royal Cornwall Gazette*, January 11, 1861
5. The East Cornwall Gunpowder Company had started producing gunpowder in a small wooded valley close to the village of Herodsfoot around 1845.
6. *The West Briton*, April 14, 1870
7. *Royal Cornwall Gazette*, February 10, 1888
8. *The Cornishman*, January 19, 1888
9. *The Hampshire Advertiser*, November 23, 1887
10. 1881 census

APPENDIX 1

1897. List of Buildings at the Bickford, Smith and Co. Ltd. Fuse works, Tuckingmill

This List is in the National Archives, UK. It was date stamped by the Home Office on 14 Oct. 1897.

The numbers on the list match a 1908 traced plan in pencil in the Cornwall Record Office.

1. Office Purpose
2. Factory Magazine for finished safety fuse.
3. Factory Magazine for finished safety fuse.
4. Lower Storey
   Taping and thread covering safety fuse and spooling tapes and yarns.
5. Upper Storey
   Coating safety fuse or instantaneous fuse with gutta percha, and measuring, cutting, sealing and tying up the same, provided that work connected with instantaneous fuse shall not be carried on at the same time as work connected with safety fuse.
6. a. Making, cutting and tying up metallic safety fuse.
   b. Adapting fuse for use in Colliery Safety Lighters.
7. Lower Storey
   Repairing and fitting work.
8. Upper Storey
   Making safety fuse or instantaneous fuse.

35. Lavatories
36. Upper Storey
    Store for Machine Patterns
37. General Store
38. General Store Shed
39. Breaking up fuses
40. Lower Storey
    Re-winding safety fuse.
41. Upper Storey
    Temporary deposit of reels of safety fuse and instantaneous fuse.
42. & 43. Miscellaneous store

9. Engine for factory.
10. Drying yarns.
11. Drying yarns and Patent Volley Firers
12. Miscellaneous Store
13. Heating Resinous Compositions by Steam Heat Only.
14. Boiler for Factory
15. <u>Lower Storey</u>
    Spooling thread
16. <u>Upper Storey</u>
    Making safety fuse.
17. <u>Lower Storey</u>
    Spooling thread.
18. <u>Upper Storey</u>
    Making safety fuse.
19. <u>Lower Storey</u>
    Making instantaneous fuse and patent Volley-Firers.
20. <u>Upper Storey</u>
    Making safety fuse.
21 & 22. <u>Lower Storey</u>
    Coalshed
23. <u>Upper Storey</u>
    Making safety fuse
24. <u>Upper Storey</u>
    Making, cutting and tying up metallic fuses.
25. <u>Lower Storey</u>
    a. Varnishing, cutting and tying up safety fuse or
    b. Varnishing, cutting and tying up instantaneous fuse, provided that the operations a. and b. shall not be carried on at the same time

44. Carpenter's work
45. a. Smithy or
    b. Testing fuse, provided that the building shall not be applied to both uses a. and b. at the same time.
46. Testing fuse and Colliery Safety-Lighters shed.
47. Boiling Tar
48. Soldering metal cases containing fuse or tubes for patent Volley-firers or Colliery Safety-Lighters
49. Miscellaneous Store
50. Miscellaneous Store
51. Miscellaneous Store
52. Keeping Instantaneous Fuse and Patent Volley-Firers.
53. Factory Magazine for finished safety fuse.
54. Factory Magazine for Colliery Safety-Lighters (cupboard).
55. Factory Magazine for finished safety fuse.
56. Packing safety fuse
57. Packing fuse
58. Office purposes
59. Keeping Methylated Spirit

26. Upper Storey
   Coating safety fuse or instantaneous fuse with gutta percha and measuring, cutting, sealing and tying up the same, provided that work connected with instantaneous fuse shall not be carried on at the same time as work connected with the safety fuse.
27. (Upper) factory magazine
   For finished fuse.
28. Keeping Pitch and Resin and Tar-Shed.
29. Work-people's lavatories
30. Lower Storey
   Work-people's dining room.

31. Upper Storey
   Work-people's dressing-room

32. Kitchen

33. Dining room

60. Keeping Tapes and Cotton and Miscellaneous Store.

61. Miscellaneous Store

62. Drying and preparing Gunpowder and Yarns for use.
63. Miscellaneous Store
64. a. Making and Drying Patent Volley-Firers
   b. Packing instantaneous fuse and Patent Volley-Firers or in lieu-
   c. Mixing Chlorate of potash and sugar
   d. Making Colliery Safety-Lighters
   e. Packing Colliery Safety-Lighters
65. a. Varnishing, cutting up and tying up safety fuse
   or
   b. Varnishing, cutting up and tying up instantaneous fuse, provided that the operations a. and b. shall not be carried on at the same time.
66. a. Varnishing, cutting and tying up safety fuse
   or
   b. Varnishing, cutting up and tying up instantaneous fuse provided that the operations a. and b. shall not be carried on at the same time.
67. Drying yarns

1908 Map to accompany Appendix 1. The numbers have been enlarged. The original traced map is in the CRO. BY/324. The map is slightly later than the list of rooms, and the fuse works had been enlarged slightly by this time. It is probable that the map was drawn when the factory was being assessed for commercial rates.

Adams Lina

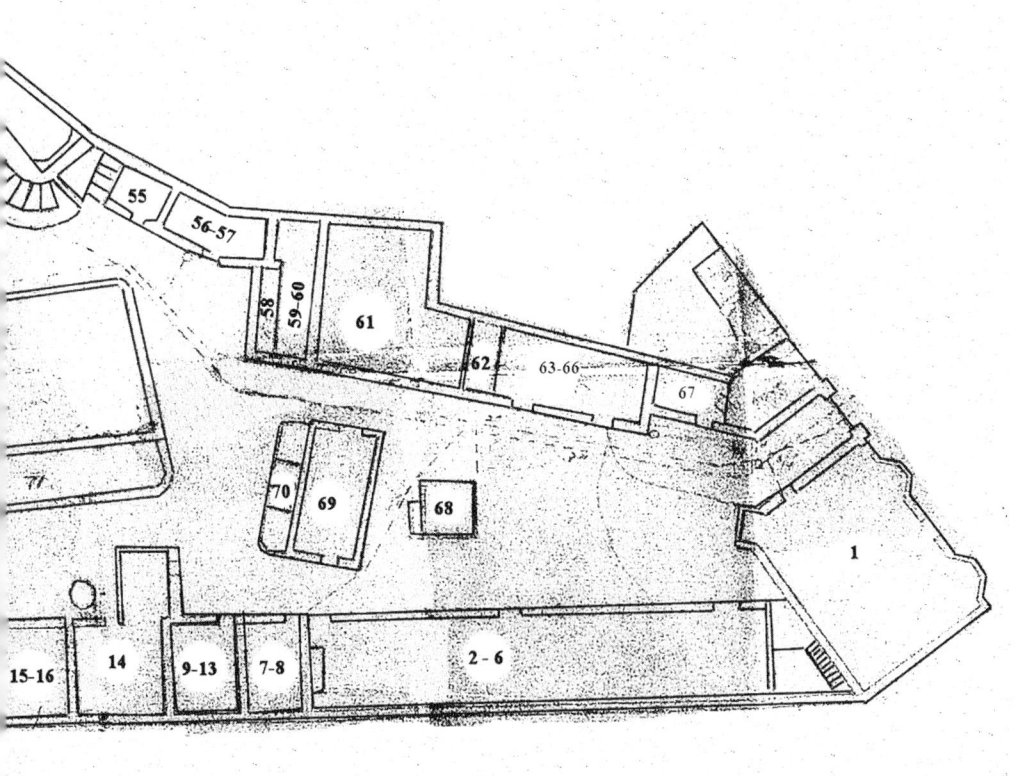

APPENDIX 2

SIGNATURES OF BICKFORD, SMITH AND CO. EMPLOYEES, JULY 1928
To mark the occasion of their long service awards at Trevarno on July 12th.
From the Peter Bickford-Smith Archive

Henry Inch	George Vincent
W. H. Inch	Percival Jobb
Laura J. Merritt	Garfield Bawden
George Lacy	Evelyn Phillips
Wilfred Beckett	Amy Laury
Leonard Philpotts	Olive Carpenter
Dorothy Paynter	Enid Wills
Maud Sincock	Hilda Wills
Dorothy Plint	Edith Mynors
Chloe Pearce	Margaret Whittle

Angove Ruby	Cock Lily	Glanville O.
Arthur Winnie	Collick Janie	George Winnie
Barbary J. Ewart T.	Collins F.	Govett Myrtle
Barbary John	Collins Mary	Gray Burnetta
Barnes Grace	Cory John	Gray Gladys
Bastion Dorothy	Crowgey Florence	Gray J.
Bawden Garfield	Curnow Gladys	Gribbin Percy
Bawden M.	Curnow M.	Griffiths Bessie
Bawden M. J.	Curry Ethel	Hampton Ada
Beckett Olive	Curry Maud	Hancock Annie
Beckett Wilfred	Daddow Annie	Hancock Gladys
Bennett E. J.	Daddow Gladys	Hanpton Lily
Bennetts A.	Daddow L.	Harris Audrey
Bennetts Beatrice	Davies Susie	Harris Beatrice
Bennetts F. W.	Davis Francis M.	Harris Bertha
Berriman Phyllis	Davis Louis	Harry T.
Bersey Emmie	Davis Myrtle	Harvey L.
Bingley Edith	Drew Liley	Harvey Lillie
Bingley Ella	Dunn Iris P.	Harvey M.
Boase Violet	Dunstan S.A.	Head Cecilie
Bodilly Carrie	Eddy Beatrice	Head Ruby
Boot V. Irene	Ellis F .M.	Hendy Viola
Bray Amy	Edwards Gwen	Henley Z. M.
Bray Ellie	Edwards Patience	Hichens May
Bryant Sam	Edwards R.	Hichens W.
Buddle Stephen	Emmett Chrissie	Hill Gladys
Burns Gladys	Endean F.	Hill Lillian
Burrall John	Eustice Julia	Hill Phillis
Burrey Mildred	Eva Annie	Hocking Ada
Bussey Lennard	Eva Hettie	Hocking Annie
Bussy James	Evans Millicent	Hocking Eva
Cady Ivy	Everson Hilda	Hollow Leonard
Carlyon Edna	Finn Ellin	Hollow W. J
Carpenter Annie	Fray Sidney	Holman Hilda
Carpenter Olive	George Eleanor	Inch Henry
Charles Maggie	Gerrard Annie	Inch W. H.
Cheffers Stanley	Gerrard Ivy	Ivey Thomas H.
Chegwidden W.	Gerrard Louie	Jackson Doris
Chubb F. G.	Gilbert Dorcas	James Millicent
Clarke Janie	Gillett W. G.	James Nellie
Clarke Mary	Glanville Clarice	James Thos.

Jeffree Freda
Jeffree Lylie
Jenkin Edith
Jenkin Nora
Jenkins G.
Jilbert L.
Johns Frank
Johns Louie
Jolly Paul
Jones Gertrude
Jose Iris
Jose Kathleen
Judd Amelia
Julian J
Julian John
Kemp Gwen
Kemp May
Kemp Winnie
Knuckey T. J.
Laity Charlotte
Laity George
Laity Lilian M.
Lawrey Salome
Lawry Amy
Lawry Florrie
Lawry Leo
Lawry Mildred
Lean J. W.
Letcher Ethel
Letcher Ruth
Lewis A.
Liddicoat D.
Liddicoat May
Lidgey Rita
Lobb A. B.
Lobb C. B.
Lobb L. M.
Lobb Percival
Lobb V. M.
Lowry Edna
Lowry Iris

Lowry May
Loye Dorothy
Luke Theo
Mead Agnes
Merritt Laura J.
Michell Daisy
Michell Doloris
Michell Violet
Mill Hilda A.
Mitchell Ruby
Moon Elsie
Moon Mildred
Moon Ruby
Moyle Henry
Myners Edith
Negus J. R.
Negus Lillie
Nicholas R. V.
Nicholls Muriel
Nile Kathleen
Painter Dorothy
Parsons R.
Pascoe Doris
Pascoe F. V.
Pascoe Millicent
Pascoe Ruth
Pearce Chloe
Pearce Mary
Pendray Lucy
Penprase Liley
Perry Doris
Phillips Burnetta
Phillips Evelyn
Phillips W. E.
Philpotts Leonard
Plint Dorothy
Polkinghorne Amy
Polkinghorne Janie
Polkinghorne Lilian
Polkinghorne Lily
Pope Evelyn

Prisk W. D.
Pryor E.
Pryor Gladys
Quintrell May
Randle William
Rashleigh A. J.
Read Phyllis
Reed Katie
Retallack Elsie
Retallack Audrey
Richards A.
Richards Hilda
Rickard Gwen
Rickard Gwennie
Roach Emily J.
Roberts Nicholas John
Roberts W. J.
Roberts Wilhelmina
Robinson H. J.
Roscholar W.
Rowe George
Rowe Gwen
Rowe Lillian
Rowe Mavis
Rowe P.
Rowe Richard H.
Rowe Stanley
Rowett Phyllis
Rule Emily
Rule Lillie
Rule M.
Rule Minnie
Rundle B.
Rundle Gertrude
Sanders Mary
Sanders Nellie
Selwood Emmie
Selwood Lenora
Selwood Linda
Selwood Myrtle
Sherman Lily

Simmons Annie	Uren Maud
Simmons Millie	Urquehart Dora
Simmons Winnie	Veale Wm. J.
Sincock Maud	Venner Alfred
Skewes Florence May	Vincent George
Skewes Katie	Visick Louie
Smith Edith A.	Walker H .L.
Smitham William	Wallace D.
Sparks Joyce	Wallace Florrie
Stapleton Elizabeth	Wallace Phyllis
Stevens Solomon	Warren Evelyn
Stoddern Evelyn	Warren P.
Stone Matthew	Waters James F.
Taylor E. M.	Wearne Annie J
Temby Ellen	Webb Gwennie
Terrill Janie	Webb Lena
Thomas Beatrice	Wells Lizzie
Thomas Edward J. C.	Whellan Augustus
Thomas Ethel	Whitford Charlie
Thomas Lillie	Whittle Ellen
Thomas Mabel	Whittle Margaret
Thow James	Williams Annie J.
Thow Jas. S.	Williams Bessie
Tonkin Bert	Williams Elsie
Treglown Ruby	Williams Hilda
Treloar Edith	Williams Janie
Trembath Laura	Williams Janie
Trescothick E.	Williams Laura
Trescothick H.	Williams Leslie
Tresidder Phyllis	Williams Lillie
Trethowan N.	Williams Violet
Trevarthan K.	Wills Enid
Trevarthen Janey	Wills Hilda
Truscott Myrtle	Wills Linda
Trythall Mabel	Wills Maud
Tyack Charles E.	Wills Queenie
Tyack Howard R.	Yeoman Ivy
Uren Ada	
Uren Fred	
Uren Josephine E.	
Uren Lottie	

APPENDIX 3

Some pages from William Bickford's patent 1831. No. 6159.
Reproduced by permission of the British Library.

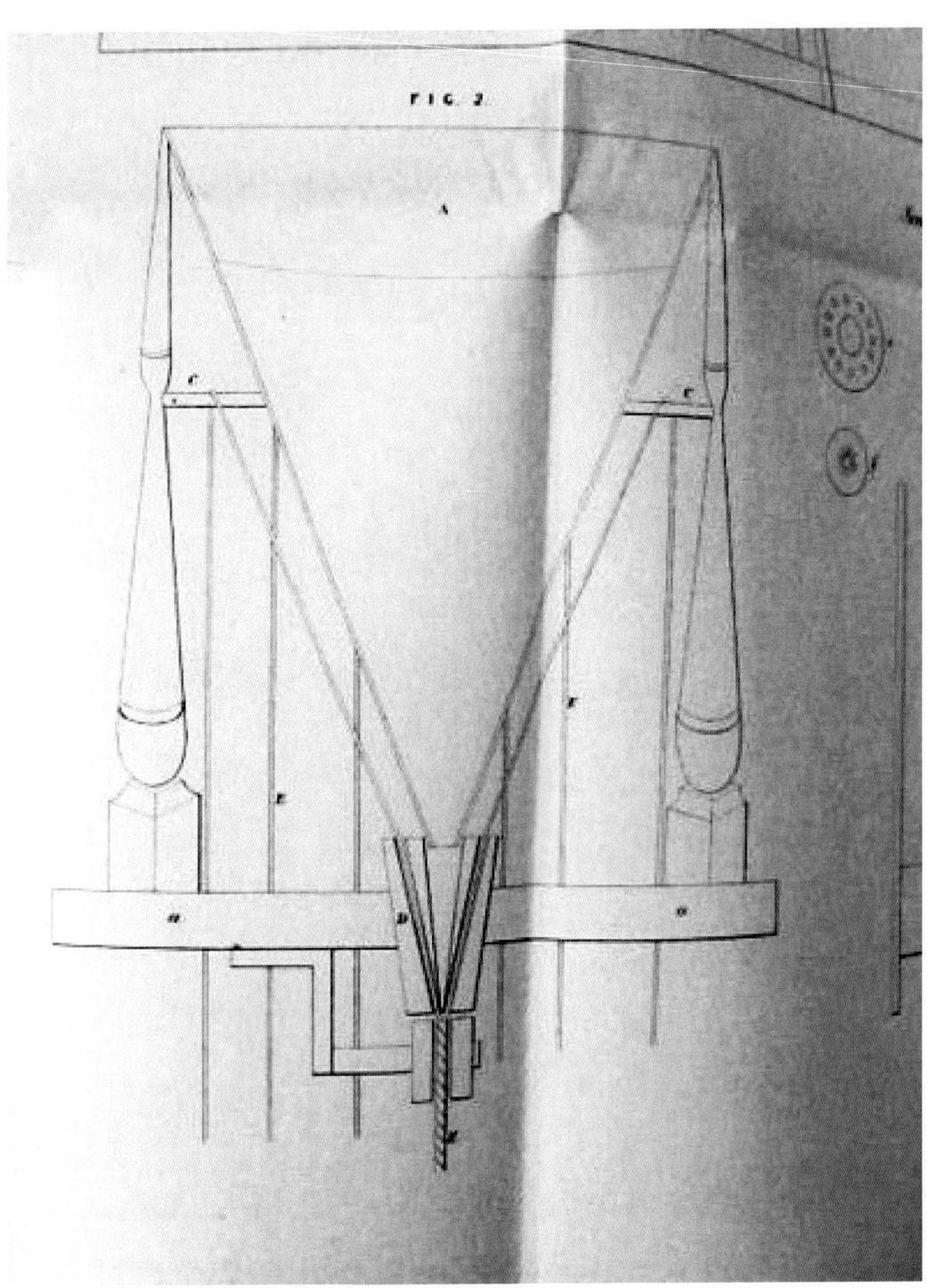

William Bickford's patent, 1831. Drawing 1. Fig. 2. The funnel through which the gunpowder was fed.

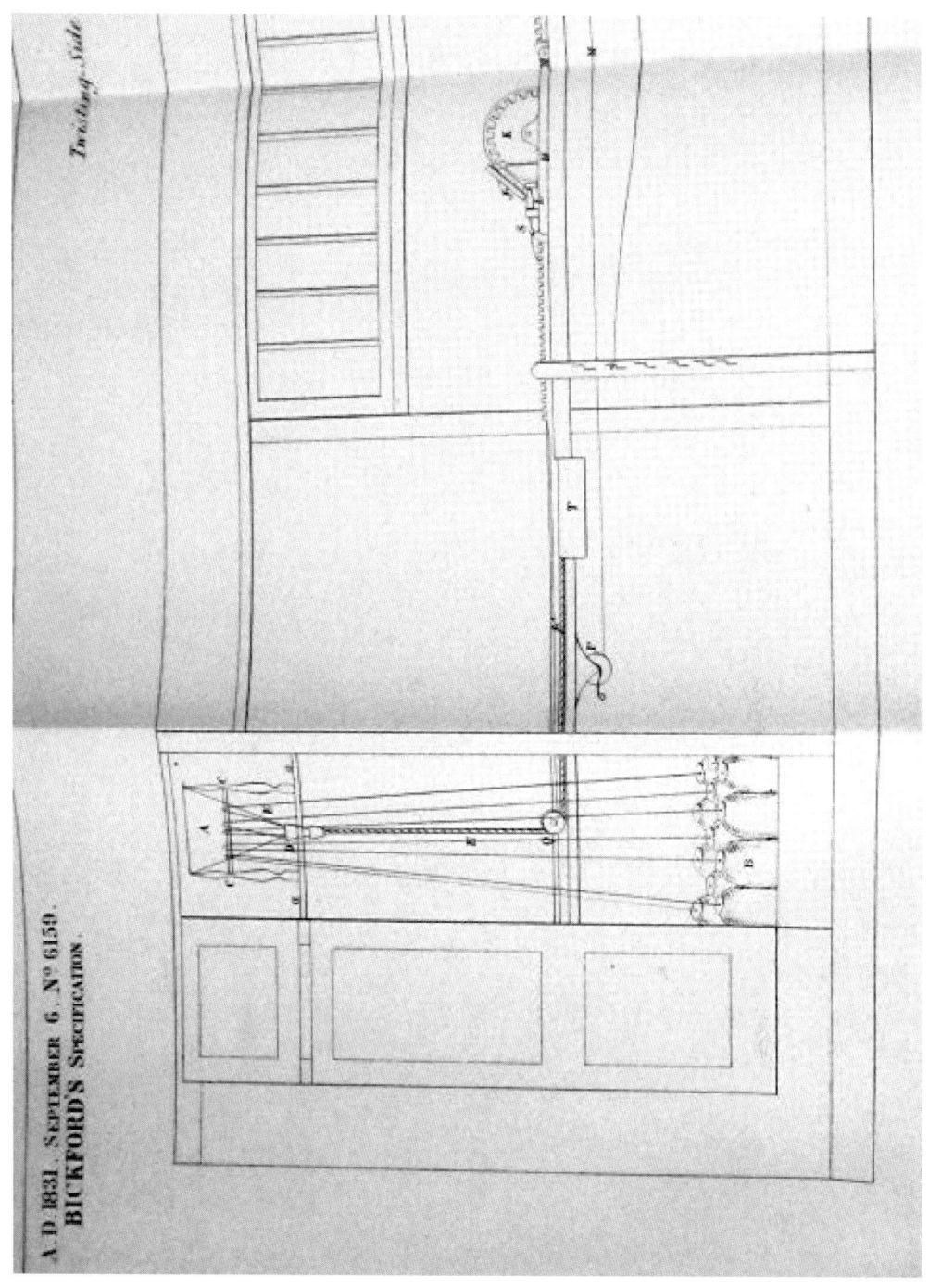

Bickford's patent. 1831. Twisting side of room. 65 feet long. The funnel is top left and the 12 spools are below. The cord is being twisted by a hand turned wheel (r)

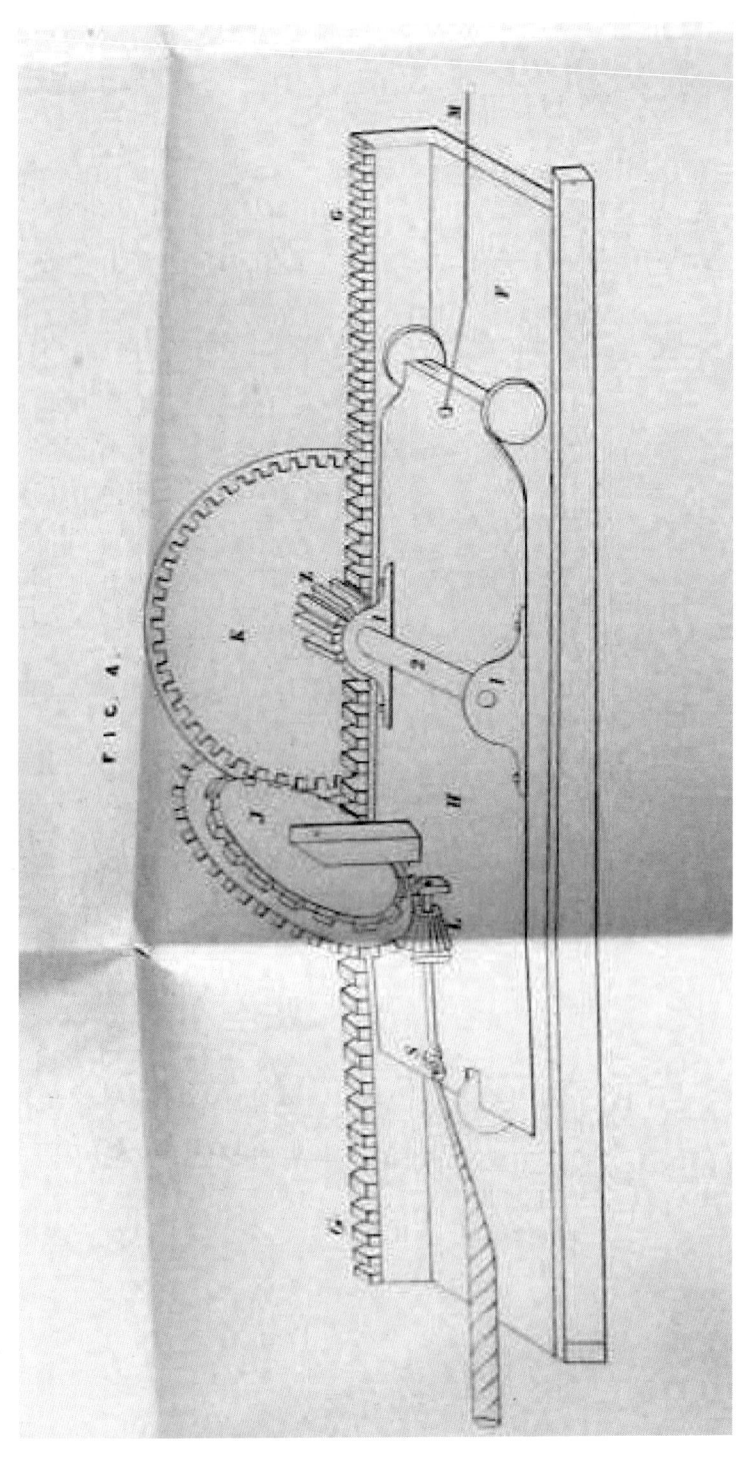

Bickford's patent 1831. Drawing of the twisting mechanism.

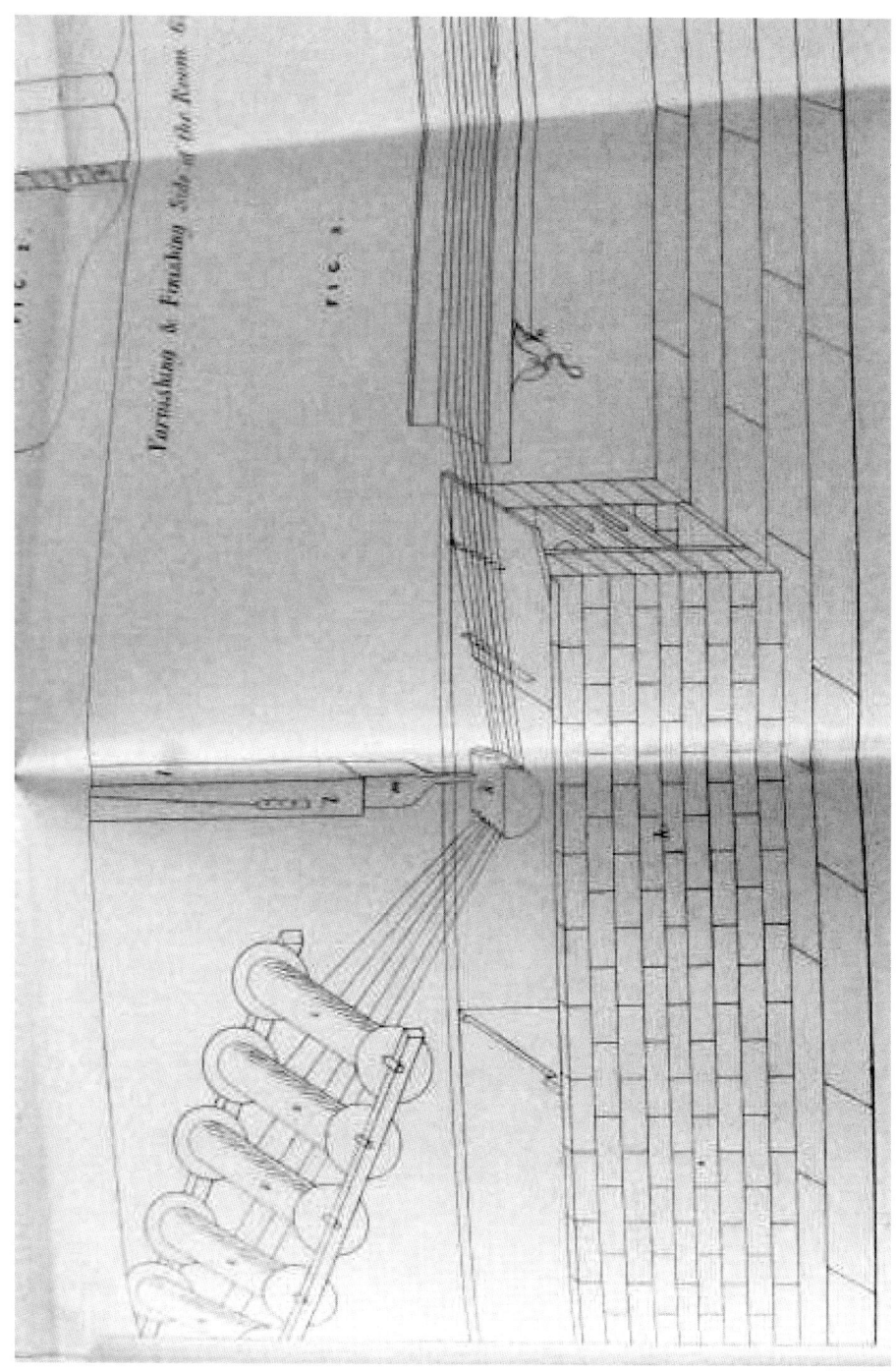

Bickford's patent 1831. The Varnishing and Finishing side of room. 65 feet long.

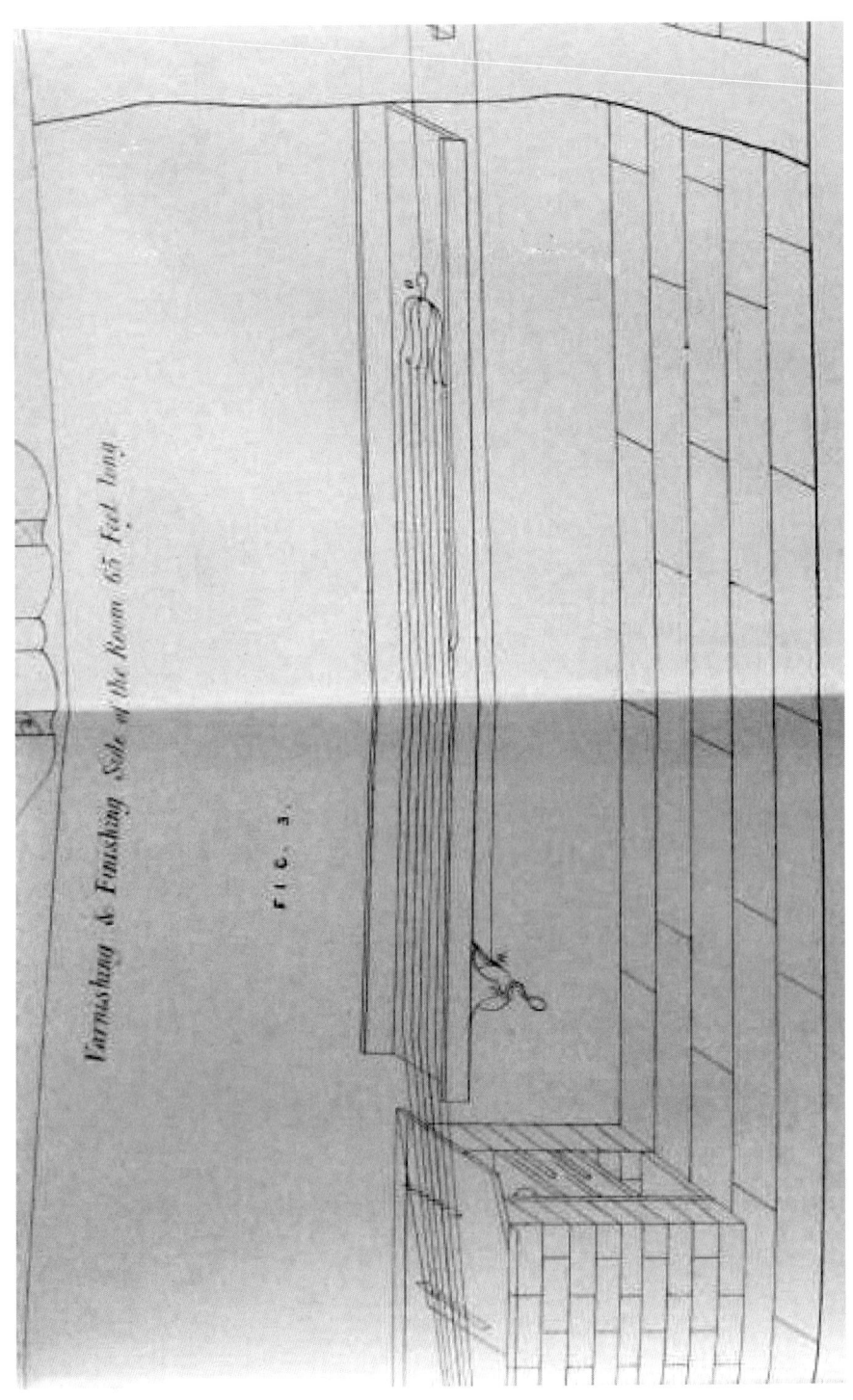

The Finishing shelf or trough, where whitening, sand or bran was applied.

Bickford's patent, 1831. Drawing 3. A perspective of the workshop where the twisting, charging and countering the fuses are in operation. 65 feet long.

INDEX

Abraham's Hotel 44, 47
Accidents in Mines, Select Committee of the House of Commons on 25
Adams, Rev. A. 158
Agar, C. B. 7
Antron Lodge 54
Aviary Cottage 206
Bacon, Bickford, Eales and Company 30
Bacon, Richard 30
Bailey, William 52
Ballantyne, R. M. 3
Barbary, John E. 111, 115
Basset family 9
Basset, Francis 6
Basset, Lady 11
Bawden, Miss A. 52
Bayly, Elizabeth Jane 148
Beacon Hill 27, 44
Beeching, Dr Richard 120
Bendigo, factory acquired in 67
Bendigo Fuse Factory 124
Bennett, Catherine 159
Bennett, Charles Frederick 159
Bennett, Edward John 159
  becomes director of Bickford, Smith and Co. Ltd. 161
Bennett, E. J., death of 80
Bennett, Ida 159
Bennetts, Fanny 49
Bennett's Fuse Works 109
Bennetts, Matthew 52
Bennett, William 34, 35, 52, 153, 159, 198, 200
  leaves Tuckingmill factory 55
  patent application 155
  patent with William Bickford Smith 153
Bennett, William and Sons Ltd. 80
Bennett, William Francis 159
  becomes manager at Roskear 161
  death of 164

Bennett, William John 163
Bickford and Davey 20
Bickford, Bacon and Co. 27
Bickford-Cordeau-Detonant 85, 87
Bickford, Elizabeth 7
  marrried 20
Bickford, John Solomon 7, 24, 34, 37, 38, 43, 44, 51, 153
  death of 43
  remarried 39
  sale of properties 44
Bickford, John Solomon Vivian 43
  death of 91
Bickford, Sarah 34
Bickford, Smith and Co. 39, 53, 59, 61
  annual outing 81
  becomes limited liability company 67
  directors of 82
  price list 57
Bickford, Smith and Co. Ltd.
  amalgamated with Bennett's fuse works 80
  annual outing 70, 75
  presentation ceremony 105
  staff 100, 106, 107, 108, 110
  visit by Prince George 99
  voluntary liquidation 111
  welfare hut 108
Bickford, Smith and Davey 20, 25, 26
Bickford, Smith and Davey and Co. 28
Bickford, Smith, Davey and Pryor 34
Bickford-Smith, George Percy
  application to start factory at Trevarno 71
  killed in Boer War 72
Bickford-Smith, Jack
  killed in accident 114
Bickford-Smith, John Clifford 74, 101, 102
  death of 115
Bickford-Smith, Michael 118
Bickford-Smith, William 66, 67, 71, 72
  See also, Smith, William Bickford

232

Bickford-Smith, William Noel 74, 101, 102
  death of 114
Bickford's Patent, legal challenge to 18
Bickford's Patent 'nippers' 80
Bickford, Susanna 24
Bickford, Venning and Co. 52
Bickford, William 10, 3, 5, 7, 8, 18, 43
  death of 24
  expiry of 1831 patent 28
  granted 'fuze' patent 20
Blackwood, Dr 76
Blamey, Catherine
  death of 26
  married 26
Blamey, Eugene 26
Blamey, John 26
Blight, Elizabeth 145, 146, 147
Blight, James 147
Blight, Jane 147
Blight, Joseph 147
Blight, Mary 148
Boer War, affect on fuse-making 75
Bond, Anna Mathilda 74, 102
Boot, Miss Violet Irene 109
Bray, Francis 7
Bray, James 7
Bray, John 7
Bray, Mary 7
British and Colonial Explosives Company, Limited 180
British and Foreign Fuse Company 28, 75
  trade mark 180
British and Foreign Fuse factory 33
British and Foreign Fuse Works
  acquired by Bickford-Smith and Nobel 76
British and Foreign Safety Fuse Company 177
British and Foreign Safety Fuse Factory 203
British and Foreign Safety Fuze Company 174
  fire at 175
British Dynamite Company 75, 198
British Empire Exhibition, 1924 109
Brunel, Isambard Kingdom 140
Brunton, Charles Robert 140
Brunton's Endless Cloth 143
Brunton's Fuse Works 141
  explosion at 145, 146
Brunton, William 140, 143
  leaves Cornwall 145
Brunton, William and Co.
  dissolution of 145
Brymbo, fuse factory opened in 147
Budge, John 12
Burall, Solomon 3
Burall, Susannah 3, 11
Burrall, Grace 11
Burrall, Kitty 11
California, new factory in 31
Camborne 5
Camborne Gas Company 11, 35, 52, 153
Camborne Literary Institute 26
Cambrian Fuse Factory 147
Cann, Francis 149
Caoutchouc 20
Carah, Emily 47
Caral, Elizabeth 50
Caral, Elizabeth junior 50
Caral, Emily 49, 50
Caral, Henry 50
Caral, William 50
Carn Brea Engine Works 143
Chanu, A. 26
Chapel Road 125
Chapel Row 9
Chatten, James 142
Christian, Elias 35
Clemo, Emily 47
Climax Company 31
Climax Rock Drill Works 158
Climo/Clemo, Emily 50
Climo/Clemo, F. S. 50
Climo/Clemo, Prudence 50
Climo/Clemo, Richard 50
Coal Mines Act, 1887 62
Coal Mines Regulation Act, 1872 52
Coast Manufacturing and Supply Company 31
Cock, Captain Joseph 193
Cock, Sarah Ann 48, 49
Collins, Ada 76
Composition exploding 88
Connecticut, factory in 26
Consolidated Mines 25

Convention of British Fuse Interests 75
Cooper-Key, Major A. 158
Copper Hill Mine 30
Copperhouse Foundry 11, 142
Copper Tankard Lode 11
Cordeau Bickford Detonant 94
Cornwall Works, Birmingham 142
Daniell, Francis 145
Daniell, Susan 145
Davey, Anna 38
Davey, Anne 192
Davey, Bickford, Smith and Co. Ltd. 87
Davey, Charles 35, 53
Davey, Elizabeth Anne 192
Davey, Mary 19
Davey, Simon 26, 35, 38, 67
  1858 patent 36
  death of 68
  General Manager at Rouen 26, 34
  married 26
Davey, Thomas 8, 19, 20, 24, 26, 35, 53
  death of 35
Davey, Thomas junior 26, 35, 37, 153
  death of 38
de Maraise, Jeanne Emma Augustine Sarrasin 35
Desborough, Captain A. P. H. 76
"Directions for using the Fuze" 23
Dolcoath Avenue 12
Dolcoath Mine 6, 12, 92
Dolcoath Road 62
Dolcoath Row 9, 69, 71, 76
  lorry accident in 81
Drewitt, PC 193
Drump Road 176
Dynamite Quay 87

Eade, Mary Anne 175
Eales, Freeman 54
Eales, Joseph 26, 30
East Hill 26
East Wheal Crofty 25
Eddy, John 11
Edward Tangye's Fuse Works 203
  closure of 206
  destroyed by fire 205
Ellis, Rev. W. 148
Ellsworth, Lemuel Stoughton 31

Ensign, Bickford and Company 30, 31
Ensign-Bickford Company 31, 87
Ensign, Ralph Hart 31
Explosives Act, 1875 54, 178, 197
Explosives Trades Ltd. 89, 162
Explosive Trades Ltd.
  becomes Nobel Industries Ltd 92

Fairfield 8
Female employment in fuse works 36, 45
Footscray, Melbourne, shares acquired in factory in 68
Forbes, M. M. 118
Ford, Major 194
Fox, George 7
Fox, Henry Backhouse 192
Fox, Robert 7
Fray, William Henry 102
Fuse Cottage 177

Gaines 88
Gardiner, Dr 76
Garland, Thomas 174
Gee, Alfred 177
Genn, W. J. 191
Gilbert, Elizabeth 147
Gilbert, James 146, 147
Gilbert, Thomas 146, 147
Glubb, A. C. L. 213
Goldsworthy, Elizabeth 50
Goldsworthy, Ellen 47, 50
Goldsworthy, Joseph 50
Goldsworthy, Thomas 50
Goose quill fuses 21
Great County Adit 189
Great Exhibition, 1851 35, 143, 188
Great Exhibition, 1862 37
Great War 87
Green Lane 176
Gribble, Mary 11
Gundry, Richard 213
Gunpowder 49
Gutta percha 25
Gweek Trading Company 60, 61
Gwennap Mines 189

Hamlin, John 194
Hancock, Ann 145, 146, 147

234

Hancock, Jane 147
Hargreaves, James 12
Harlé, A. 67
Harris, John 193
Haslett, John 61
Hawke, Charlotte 191
Hawke, Edward 190
Hawke, Edward Drummond Hay 201
Hawke, Edward H.
   death of 191
Hawke, Edward Henry 187
Hawke, Edward junior 190
   death of 191
Hawke, John Anthony
   becomes King's Counsel 200
Hawke, Julia 191
Hawke, Mary 189
Hawke, Richard 211, 212
   death of 213
Hayle Railway 25, 140
Herbert, Luke 17
Henry, Edward 187
Heyden, John 180, 181, 182
Heyden, William 180, 181
Hicks, James 59, 61, 62, 177
Hitchens, Thomas 8
Hocking, John 174
Hocking, Samuel 47, 48, 54
Hodge, W. P. 194
Holman Brothers 89
Holman, Percy M. 118
Hosking, Emily 149
Houseman, Henry 213
Hunter, John 67
Hutchinson, Thomas 48, 51
ICI Explosives Division
   renamed Nobel Division 119
Imperial Chemical Industries (Explosives)
   Ltd. 110
Imperial Chemical Industries Ltd. 103
International Exhibition, Dublin 176
Irons Brothers Foundry 37
Ivey, Elizabeth 49
Jacobi, Ernest 26
Jacobi, Francis 26
James, Annie 47, 50
James, Eliz 50

James, Eliza Jane 192
James, Grace 50
James, Susan 192
Jenkin, Alfred 59, 61
Jones, Ann 175
Jones, Caroline 194
Kennall Vale 49
Killifreth Mine 190
Kingstown Harbour, underwater blast at 25
Knight, Peter 7
Lanyon, Alfred 177, 181
   granted patent 180
Lanyon, John Charles 174, 175
   death of 175
Launder, John 177
Launder, Richard 178
Launder, Sarah 175
Launder, William Henry 28, 174, 175, 177, 178
Lawry, Catherine 153
Lawry, John 153
Lean, J. W. 76, 102
Lelant Quay 87
Lemin, PC 76
Longclose Lode 11
Long, Margaretta 192, 193, 194, 197
Luke, Catherine 194
Luke, Kate 192
Majendie, Major V. D. 196
Manley, William 180
Marks, Benjamin 50
Marks, Eliza Ann 50
Marks, Elizabeth 50
Marks, Elizabeth Ann 47
Marks, Ellen 50
Marks, Thomas 50
Mason, Charlotte 187
Matthews, Mary Ann 49
May, Charles 48
Mayne, Captain 193
May, Oliver 144, 147
May, William 60
McGowan, Harry 75, 162
McGowan, Sir Harry 91, 103, 110, 200
Medals won by Bickford Smith and Co. 57
Medland, Joseph 211

Meissen, new factory in 26
　fire at 54
Melbourne Exhibition, 1881 151
Merritt, James B. 31
Michell, Christiana 192
Michell, Fanny 189
Michell, James 198
Miners, Emma 48
Miners' safety rod 21
Mitchell, George 197
Mitchell, Thomas 11
Mond, Sir Alfred 103
Morcom, Michael 174
Morwellham Quay 8
National Explosive Company Ltd. 113
Needle 22
Nettell, Mary J. 175
New Dolcoath Mine 163
New Orleans and Pontchartrain Railway 140
New Roskear Shaft 163
Newton, Llewellyn 53
Nicholls, Henry 79
No. 44 Fuze 88
Nobel, Alfred 67
Nobel Division 109
Nobel-Dynamite Trust Company 75
Nobel Industries Ltd. 163
　becomes Explosives Division of ICI 109
　Local Board 102
Nobel's Explosives Company 71
　acquires Penhellick fuse works 71
　acquires Tangye's fuse works 71
Nobel's Explosives Company Ltd. 75, 89
Nobel's Explosives Ltd. 198
North Roskear Mine 21, 24, 25
North Wheal Crofty 25

Opie, Elizabeth 210
Opie, Ellen 148, 149
Opie, Grace 214
Opie, James 214
Opie, Joseph 210, 211
Opie, Susanna 210
Opie, Thomas 214
Opie, William 149

Paris, John Ayrton 7
Parkyn, Robert 39

Patent Safety Igniter 55, 63
Pearce, Vivian 67, 151
Pendarves Estate 153
　sale of 165
Pendarves, E. W. W. 11
Pendarves House 164
Pendarves, John Stackhouse 165
Pendarves Street 9, 62, 69, 84, 127
　commemorative plaque 103
Pendarves, William Cole 163
　death of 165
Pengegon
　jute spinning factory in 44
Pengegon steel works 69
Penhellick 34
Penhellick Fuse Works. *See also* Brunton's Fuse Works
　closure of 151
　explosion at 147
Penlu
　enlargement of 43
　sale of 38
Permitted Igniter Fuse 79
Perry, Charles 67
Peter, William John 160
Phillips, John 148
Phoenix Mining Company 30
Phoenix Ropewalk 35
Pike, Robert Hart 51, 141, 143, 144
Pike, Walter 51, 141, 145
Pinwell, Charlotte 187
Pinwell, John 187
Pinwell, Julia 187
Plain-an-Gwarry 34, 176
Plint, Martha 48
Poldice Mine 189
Poldice Mine, West 193
Polglase, James 37
Pooley, Elijah 193
Pooley, Elizabeth Anne 192
Pooley, John 194
Popham, G. W. 53
Porthleven Institute 62, 124
Pryor, Captain William 40
Pryor, Francis 34, 39

Rabling, W. 60
Raleigh Cycle Company 88

Redbrook House 44
Red River Canal 140
Redruth and Chasewater Railway 140
Redruth Brewery 43, 174
Relistian Mine 24
Renfree, Alfred 147
Renfree, Herbert 147
Renfree, Peter Alfred 144-149
  death of 151
Rice, Henry 211
Richards, Emily 47
Richards, Mary Ellen 76
Rich, William
  death of 199
Rifle Volunteer Corps 51
Rodda, Elizabeth May 76
Roscroggan Free United Methodist Chapel 115
  used as store 89, 114
Roseworthy Hammer Mills 11
Roskear Fuse Works 153, 170, 171
  breaches of Regulations at 156
  Crown Brand 156
  fire at 158
Roskear Row 154
Roskear Villas 52
Rouen Company, partners in 26
Rouen, new factory in 26
Rounsevell, William 37
Rowe, Mary 175
Rowe, Sarah 174
Rowe, William 157
Royal Geological Society, Cornwall 7
Safety Bar 7
Safety fuse
  burning rates of 21
  failure rate of 24
  lengths of 21
  lengths sold in 23
  manufactured lengths of 23
  price of 21
  production in 1895 69
  production of 84
Safety Rod 7
Saltpeter 49
Scinde Railway Company 145
Simms, Ellen 49

Sims, Ellen 48, 50
Sims, Henry 50
Sims, Louisa 50
Sims, Louisa Ann 47
Sims, Mary 50
Sims, Mary Ellen 47
Sims, William 50
Skewes, the Younger 18
Sleeman, Mrs 43
Sleeman, William 145, 146, 147
Smith and Co. Ltd. Sports Pavilion 99
Smith, Colonel George Edward Stanley 74, 102, 104, 109
Smith, Dr George 9, 18, 27, 190
Smith, Elizabeth Burall 24, 44, 66
  death of 62
Smith, George 24, 26, 48, 143
  appointed Chairman of the Cornwall Railway 34
  death of 39
  marriied 20
  purchases old Camborne Workhouse 27
  trustee of Camborne Wesleyan Chapel 26
  writing religious pamphlets 26
Smith, George John 39, 44, 55, 57, 61, 67
Smith, George junior 27
Smith, G. E. Stanley 199
  war service 87
Smith, Harold O. 87, 102
  becomes Chairman of Board of ICI Metals Division 109
Smith, Henry Arthur 44, 67
Smith, Sir George John 8, 66, 70, 74, 80, 91, 199, 200
  death of 92, 163
Smith, William Bickford 35, 39, 41, 44, 51, 55
  See also, Bickford-Smith, William
  becomes Liberal MP 62
  changes name to Bickford-Smith 62
  opens Porthleven Institute 62
  purchases Trevarno 53
South Caradon Mine 212
South Roskear Mine 20, 155, 163
South Roskear Terrace 164
Spargoe, Edmond 7
Sparnon, Catherine 182

Sports Pavilion 126
Spotswood, Melbourne, new factory in 68
Sprague, Lily 76
St. Day Fire Brick and Clay Company 200
Stephens, Cecilia 192
Stephens, Rosella 193
Stephens, William 158
Stray Park Mine 6
Swansea Safety Fuse Company 59, 178, 206
Swansea, Wales Swansea Fuse Co. 71

Tangye, Alfred 203
Tangye, Anne 203
Tangye, Edward 180, 203
  patent for tube-rolling machine 203
  patent fuse 203
Tangye, Ernest 203
Tangye, George 203
Tangye, James 142, 203, 205
Tangye, Joseph 142, 203
Tangye, Lancelot 204
Tangye, Richard 142
Tangye, Sarah 203
Taylor, John 25
Teague, Charles Dysart 199
Teague, Richard Henry 199
Tehidy Estate 8
Tetryl 88, 93
  side effects of handling 161
The Elms 176, 203, 204
Thomas, James Henry 181, 210
Thomas, William 210
Tincroft Mine 141
Tolgarrick Road 25
Tolgullow Farm 191
Tolvean House, Redruth 176
Towan, Eliza 50
Towan, Martha 47, 50
Towan, Mary 50
Towan, Sarah Ann 50
Toy, Ann Hosking 31
Toy, Bickford and Company 31
Toy, Joseph 30, 31
Toy, Robert 31
Tregonning, James 194, 196, 197
Tregothnan Consolidated Mines 26
Trelawny Pierrot Troupe `04, 162
Tremar Coombe 210

Trengove 8
Trevarno 55
Trevarno, California 31
Trevarton, Annie 48
Trevithick, Richard 3, 6
Trevu 27, 44
Trewartha, Elijah 198
Trewartha, Elisha 191
Trinitrophenyl methylnitramine 88
Trinitrotoluene 86, 93
Tuckingmill 9
Tuckingmill factory 42, 95, 96, 98
  enlarged 59
  expansion of 34
  fire at 45, 46, 47, 49, 51, 76, 78
Tuckingmill Foundry 11, 43, 69, 81
  demolition of 84
  fire at 75
Tuckingmill Foundry and Rock Drill Company 69
Tuckingmill Hotel 43
Tyack, Charles Edward 60, 67, 74, 81, 102
  death of 109
Tyack, J. 60

UK Safety Fuse Trade Association 80
United Mines 25
Unity Fuse Company 28, 34, 162, 199
  explosion at 188
  fire at 192
  purchased by William Rich 198
Unity Patent Safety Fuse Company 198, 199
  fire at 196
Upton Towans 142
USA import duty on fuses 30

Vanguard Rock, Plymouth 36
Vaughan, Pearce 175
Venning, Margaret Leaman 35
Victor, Joseph 37
Vincent, Elizabeth Jane 49
Vivian, Captain Andrew 3, 6
Vivian, Edward 34
Vivian, Elizabeth 148, 149
Vivian, Grace 149
Vivian, Henry Andrew 39
Vivian, James 102
Vivian, John 7
Vivian, Julia 39

Vivian, William 11, 34, 149
Volley firer. *See also* Patent Safety Igniter
Vyvyan, Sir R. R. 54
Watson, James 67
West Cornwall Railway 140, 155
West Wheal Towan, gunpowder mill and magazine at 37
W. G. Armstrong Whitworth and Co. Ltd. 88
Wheal Gons 6
Wheal Plosh 11
Wheal Susan 11
Wheal Unity Wood 189
Wiener Neustadt, factory site acquired in 54
Wilkinson, Mence 177
Willacy, J. A. 118
Williams, Annie Jane 156
Williams, Jessie 76
Williams, John 7
Williams, Mary 156
Williams, Michael 188
Williams, Sir Frederick 192, 193, 198
Women's Hospital, Redruth 76
Yorks 24